Scottish Wildlife

Animals

First Published in Great Britain by
Colin Baxter Photography Ltd.,
Unit 2/3, Block 6,
Caldwellside Industrial Estate,
LANARK, ML11 6SR

British Library Cataloguing in Publication Data
Collier, Ray
    Scottish wildlife : animals
    I. Title
    591.9411

    ISBN 0-948661-21-6

Photography by:

Front cover © Laurie Campbell (NHPA)
Back cover © Laurie Campbell
Page   4 © Laurie Campbell
Page   6 © Dennis Bright (Swift)
Page   8 © Hans Reinhard (Bruce Coleman)
Page   9 © Mick Chesworth (Aquila)
Page 10 © Manfred Danegger (NHPA)
Page 11 © Manfred Danegger (NHPA)
Page 12 © Laurie Campbell
Page 13 © W S Paton (Bruce Coleman)
Page 14 © Gordon Langsbury (Bruce Coleman)
Page 15 © Wayne Lankinen (Bruce Coleman)
Page 16 © Rod Williams (Bruce Coleman)
Page 17 © Robin Redfern (OSF)
Page 18 © Wayne Lankinen (Aquila)
Page 19 © Stephen Dalton (NHPA)
Page 20 © Laurie Campbell (NHPA)
Page 22 © Laurie Campbell
Page 24 © Mike Read (Swift)
Page 25 © W S Paton (NHPA)
Page 26 © Laurie Campbell
Page 27 © Laurie Campbell
Page 28 © W S Paton (NHPA)
Page 30 © W S Paton (Nature Photographers Ltd)
Page 31 © Hans Reinhard (Bruce Coleman)
Page 32 © Laurie Campbell
Page 34 © Laurie Campbell
Page 35 © Laurie Campbell
Page 36 © George McCarthy (Bruce Coleman)
Page 38 © E A Janes (Nature Photographers Ltd)
Page 39 © George McCarthy (Bruce Coleman)
Page 40 © Neil McIntyre (Swift)
Page 41 © John Daniels (Ardea)
Page 42 © Laurie Campbell
Page 43 © W S Paton (Bruce Coleman)
Page 44 © Adrian Davies (Bruce Coleman)
Page 46 © Laurie Campbell
Page 47 © Laurie Campbell
Page 48 © Laurie Campbell
Page 50 © Paul Taylor (OSF)
Page 51 © Jean-Paul Ferrero (Ardea)
Page 52 © E A James (NHPA)
Page 54 © Mike Read (Swift)
Page 56 © Mike Wilkes (Aquila)

Page 57 © Keith Ringland
Page 58 © Gordon Langsbury (Bruce Coleman)
Page 59 © Keith Ringland
Page 60 © Laurie Campbell
Page 61 © Laurie Campbell
Page 62 (Top) © Laurie Campbell
Page 62 (Bottom) © H D Brandl (FLPA)
Page 63 © Gordon Langsbury (Bruce Coleman)
Page 64 © George Bernard (NHPA)
Page 65 © Laurie Campbell
Page 66 © Stephen Dalton (NHPA)
Page 67 © George McCarthy (Bruce Coleman)
Page 68 © Stephen Dalton (NHPA)
Page 69 © Andy Purcell (Bruce Coleman)
Page 70 © Paul Sterry (Nature Photographers Ltd)
Page 71 © George McCarthy (Bruce Coleman)
Page 72 (Top) © Stephen Dalton (NHPA)
Page 72 (Bottom) © Laurie Campbell (NHPA)
Page 73 © John Markham (Bruce Coleman)
Page 74 © Rod Williams (Bruce Coleman)
Page 75 © Stephen Dalton (NHPA)
Page 76 (Top) © Stephen Dalton (NHPA)
Page 76 (Bottom) © Jane Burton (Bruce Coleman)
Page 77 © Paul Sterry (Nature Photographers Ltd)
Page 78 © Andy Purcell (Bruce Coleman)
Page 79 © Rod Williams (Bruce Coleman)
Page 80 © Laurie Campbell
Page 81 © Laurie Campbell
Page 82 © M P L Fogden (Bruce Coleman)
Page 83 © P Morris (Ardea)
Page 84 © W S Paton (Bruce Coleman)
Page 86 © Laurie Campbell
Page 87 © E A Janes (Nature Photographers Ltd)
Page 88 (Top) © M B Withers (FLPA)
Page 88 (Bottom) © George McCarthy (Bruce Coleman)
Page 89 © Stephen Dalton (NHPA)
Page 90 © Dick J C Klees (Bruce Coleman)
Page 91 © David Woodfall (NHPA)
Page 92 (Top) © Paul Sterry (Nature Photographers Ltd)
Page 92 (Bottom) © George McCarthy (Bruce Coleman)
Page 93 © Paul Stevens (Swift)
Page 94 © Keith Ringland
Page 95 © George McCarthy (Bruce Coleman)

Printed in Hong Kong

Scottish Wildlife

Animals

by

Ray Collier

Colin Baxter Photography Ltd, Lanark, Scotland

CONTENTS

CARNIVORES

Fox	8
Pine Marten	10
Stoat	14
Weasel	16
Mink	18
Polecat	19
Badger	20
Otter	22
Wildcat	28

UNGULATES

Red Deer	32
Sika Deer	36
Fallow Deer	38
Roe Deer	40
Reindeer	42
Wild Goats	44
Soay Sheep	48

RABBITS & HARES

Rabbit	50
Brown Hare	52
Mountain Hare	54

RODENTS & INSECTIVORES

Red Squirrel	58
Grey Squirrel	64
Hedgehog	66
Mole	68

SMALL MAMMALS   70

Wood Mouse
House Mouse
Harvest Mouse
Water Vole
Bank Vole
Field Vole
Common Shrew
Pygmy Shrew
Water Shrew
Black Rat
Brown Rat

SEALS

Common Seal	80
Grey Seal	84

AMPHIBIANS & REPTILES   88

Great Crested Newt
Smooth Newt
Palmate Newt
Common Toad
Natterjack Toad
Frog
Slow-worm
Common Lizard
Adder

Bibliography	96

INTRODUCTION

'The strath was brooded over by that magical High-lands' silence in November and I had just been watching three species of deer – red, sika and roe – in a woodland glade together. Further up the strath that reached into the Monadhliath hills I watched a stoat, a female by the size of it, running purposefully. She crossed an area of short heather and then ran onto a large expanse of scree, where six mountain hares in their white winter coats were spaced out. At first I thought the stoat was after one of the hares, but although she passed within two feet of one of them, she ignored it completely. The last I saw of the stoat was as she ran up a small, but steep, rock face and went over the top, out of sight. All this was also viewed by a large black billy goat with huge curved horns. As he surveyed the scene, he sniffed at the breeze which was wafting over the scent of a group of nannies further along the steep slopes, no doubt reminding him of the rut which had just ended.'

The above notes were written following one of my many visits to Strathdearn, a few miles south of Inverness. The aim of this book is to share the pleasures that Scottish wildlife has given me in the last 30 years, from the spectacular seal rookeries on the Monach Isles to roe deer in woodland and from black water voles to great crested newts in old curling ponds.

Throughout my varying contacts and work with wildlife in Scotland the rule has always been that 'watching wildlife is fun, a science often, but always fun'. Unlike birds and plants very few people have studied the whole range of species of mammals and amphibians and reptiles mentioned in this book. With some species such as the Soay sheep, red deer and seals there has been a great deal of general observation and research. For others, there have been just a few brief observations.

Scottish Wildlife – Animals is an introduction to the rich variety of animals found in Scotland today, with up-to-date notes on their behaviour, distribution, persecution and conservation issues. One of the reasons that many people have not taken an interest in our wildlife is the difficulty of even seeing, say, a pine marten, badger or perhaps roe deer in the wild. This book aims to help you discover and enjoy wildlife and there are suggestions in the text for making contact with many of our resident species.

There is no longer any completely untouched wilderness in Scotland, as even the high tops and the globally important Flow country have been modified in some way by man. But the quotation from Henry Thoreau still rings true for the wildlands of Scotland today: 'In wildness is the preservation of the world.'

Ray Collier

FOX

There are probably more stories about the cunning nature of the fox and the means by which it eludes the hunt and catches its prey than about any other mammal in Scotland. However, to see one walking on a hillside in its winter coat, when it is full and thick, one would think that it had an air of entire innocence. In actual fact its reputation as a predator, and it is the most widespread and abundant one all over Britain, is well deserved, whether it be in a remote glen in the Highlands or in a suburban back garden. It has been calculated that an average of 50,000 foxes are killed in Britain every year by hunting and shooting and yet incredibly it seems to make no difference to the population. As with other mammals, such as the grey squirrel, it may well be that by killing them in one area a vacuum is created that other foxes can fill.

The presence of foxes is most evident during the mating period from early January to around February, when during darkness the triple bark of the dog fox can be heard, sometimes followed by the unearthly answering scream of the vixen, shattering the night air. At this time of year foxes are occasionally seen in daylight, but for the rest of the year they are mainly nocturnal. However, anyone out at dusk, particularly at the time of short summer nights, stands a good chance of seeing them returning to their den or retreat. This is not necessarily the traditional hole in the ground as foxes will also hide in a dense thicket or even up a tree. They are agile climbers and often choose a pollarded willow.

The ingenuity of the fox is only partially matched by the ingenuity of its main predator – man. Shooting at night relies on the fact that the fox can be attracted by the imitation of a rabbit screaming, which it thinks is an injured animal and a food source. A strong spotlight is then reflected in the eyes of the fox and a rifle is used to finish it off. Foxes are often shot at first light when they are going back to the den with food for the cubs. Fox snares are still legal providing they are not self-locking and are inspected every day. The fact that long dead foxes or, for that matter, badgers and pine martens are found in such snares, is a ready indication that the law is not respected in many areas.

Following the mating season, the single litter of three to eight cubs is born in the underground den or earth in March or April. When the cubs are newly born the vixen will stay with them whilst the dog brings her food. The cubs are round-faced and short-eared, a little like puppies, and are covered with dark chocolate fur which is replaced by a reddish-brown coat after eight weeks. The cubs first appear above ground after 24 days. Non-breeding vixens may help to feed the cubs as most foxes live in family groups, consisting of the dog, vixen, cubs and, perhaps, one or two non-breeding vixens. The young dogs, however, leave the family group at the end of the year to set up territories of their own.

Foxes are increasingly found in urban gardens where some people enjoy their presence and others object strongly. If left alone they will sometimes even breed in gardens, with the den being under the garden shed, behind a crate or the fox might even dig a den out. At night the foxes forage over a large area and will regularly take lids off dustbins in search of food or raid bird tables in the winter. Cats and foxes are wary of each other and generally leave each other alone, although when confrontations do take place the fox is generally the first to back away.

PINE MARTEN

Cliffs and scree associated with woodland are the main haunts of this attractive, mainly nocturnal and elusive animal. Most accounts of the pine marten's habits stress its affinity with woodland and paint a picture of the animals constantly chasing through the trees after their prey, but in practice they can spend as much time on the ground. Fortunately, although it is an elusive animal, its presence is easily identified by its droppings. They are small, dark and twisted and once seen, never forgotten. It is surprising just how many can be found in some areas.

Identification is not as easy as some books suggest and proof of this is that about half of the dead 'martens' that are reported, have, on investigation, turned out to be polecat ferrets. These have almost certainly escaped or been deliberately released by people who used them for catching rabbits. The best feature for identification of the marten is the general dark brown of its fur that contrasts with the creamy throat patch, but its bounding gait and bushy tail are also a help. It is one of the few predators that is agile enough to catch squirrels and it has been introduced to some parts of southern Scotland to try, usually with little success, to keep the number of grey squirrels down.

In the 17th century it was called a matrix or mertrich and persecution by man has always been unusually severe. Perhaps only the badger and possibly the wildcat have been persecuted with quite so much vigour and the often quoted 'vermin' list from Glengarry is most revealing. The following animals were killed on this estate, south-west of Inverness, between 1837 and 1840: '11 Foxes, 198 Wildcats, 78 House cats, 246 Pine Martens, 106 Polecats, 301 Weasels and Stoats, 67 Badgers'.

The infamous gin trap took its toll and it is a sobering thought that this deadly trap was made illegal in England in 1954, whilst deplorably, it was not until 1973 that a similar move was made in Scotland. Now the pine marten is fully protected by the Wildlife and Countryside Act 1981, although persecution undoubtedly still continues as, for some people, tradition dies hard. However, it seems likely that more are now kil-

Pine martens are agile climbers and can prey on red squirrels.

led on the road by vehicles and the increase in such fatalities is an indication of its range and numbers.

In the last ten years or so the distribution of the marten has spread, with spectacular increases taking place along the Great Glen. Here, as elsewhere in Scotland, the decrease in persecution has coincided with recent afforestation that provides a sanctuary for them, with only the occasional fox snare to contend with. The 1981 legal protection came as a surprise to many people, particularly in the Highlands where the population was increasing, but in a British and European context the animal was rare. However, as with the wildcat there was a serious anomaly in the first Act of 1981 and it was not until an amendment came in 1988 that these animals were fully protected. In the 1981 Act, an authorised person could shoot a marten or wildcat that was causing damage, but the same person could not trap it and take it away for safe release. The Amendment in 1988 covered this loophole and gave both species full protection. Most of the British colonies outside Scotland are now extinct and there is little chance of the pine marten spreading naturally from north to south Scotland because of the great

conurbations of Glasgow and Edinburgh.

Although martens have been studied by a few naturalists, there is still a great deal we do not know about them. The increase in range and numbers has brought it into conflict with man as it raids chicken houses, or even comes into the roof space of homes where it will leave remains of prey. Generally, with all but the most persistent individual, it is a question of having adequate fencing. There are also reports of pine martens in goldeneye nest boxes, which was proven when an observer put his hand into such a box expecting to find the duck and/or eggs only to find a pine marten. There is even the suggestion that they take adult Slavonian grebes but this has yet to be verified as there are a number of other animals that could be the culprit, such as fox, wildcat, mink or even otter. Pine martens will raid bird colonies, as happened a few years ago on an island in Loch Sheil, when they completely wiped out the year's brood in a colony of common gulls by eating all the chicks.

Because of their nocturnal habits, watching martens is not easy and is mostly restricted to fleeting glances in the headlights of vehicles. In the summer martens will raid litter bins, and droppings around such bins can give an indication that nightly raids will take place. If they are known to be in an area, they are very partial to toast covered with jam, or hens' eggs, and can be attracted by food left out for them. There is one logcabin-style holiday centre near Fort William, where watching martens is included as one of the attractions of the holiday. The marten's natural diet is obtained mainly on the ground and consists of small birds and mammals but also includes a wide range of items such as berries, eggs, beetles and caterpillars.

Dens can be found in a variety of places such as cairns, crags, scree and sometimes holes in trees. The breeding season is in July and August but there is a delay in the implantation of the fertilised egg. Consequently, the female does not usually become pregnant until January, and then bears an average of three young in March or April. At eight weeks' the young start to appear outside the den and although they grow rapidly, it is not until the following summer that they breed. Apart from man martens have no natural predators in Scotland, although on the Continent they are sometimes taken by golden eagles.

The white winter coat (Opposite) is called 'ermine'.

STOAT

One of the reasons for the success of the stoat is that given sufficient cover, it can successfully colonise a very wide range of habitats. It is found in grassland, moorland, woodland, scree, hedges, stone walls or outbuildings. Dense cover is essential to the stoat's survival given its small size. It can fall prey to a number of predators such as birds of prey and larger mammals. Both the stoat and the weasel are unusual in that the size between the males and females varies considerably. The length of the head and body of a male stoat is some 12 inches whereas a female is only about 9½ inches. Most of the stoats in Scotland turn a creamy-white colour for the winter but always retain the black tip to their tail. In southern Britain stoats do not moult into this white coat except in extremely cold winters. However they sometimes only partly change so the animal has a pattern of brown and creamy fur. The white winter coat is known as ermine and in other countries has been much sought after as fur.

Stoats are found throughout mainland Scotland but are absent from the Western Isles. Gamekeepers' gibbets these days are likely to contain mainly stoats and weasels, as many of the other mammals that formerly appeared are now protected. Stoats have always been intensely persecuted despite the fact that up until the outbreak of myxomatosis their main food was rabbit. This is demonstrated by the fact that when the rabbit population was decimated by the disease, the numbers of stoats dropped accordingly. Most prey is taken under some sort of cover but they will take animals such as rabbits in the open. To do this the stoat carefully utilises every last piece of cover before it leaps at the victim, always striking at the back of the neck and with a deep bite. Its habit of licking blood from the wound has given rise to the myth that stoats suck blood from their victims. Another ploy used by the stoat is to run very swiftly in tight circles round its intended victim – this is often called the 'stoat dance' – and the unfortunate victim becomes mesmerised until the stoat leaps for its neck. For its size, this slim but savage animal is one of the fiercest of predators. Active both day and night, it relentlessly tracks down its prey by scent. A single rabbit will be followed through a colony, whilst the other rabbits carry on feeding. Occasionally the rabbit in question will become panic-stricken and will simply lie down and squeal as the stoat approaches.

Stoats mate in summer but implantation of the fertilised egg is delayed until the following March, and the female gives birth to six or more young in April or May. The young will leave the den – generally in a hollow tree, rock crevice or burrow – after five weeks and then they often hunt and play in a family group, which has led to the idea that they hunt in packs. The young are independent after ten weeks and are small enough to be mistaken for weasels but the black tip to the tail readily distinguishes them. Stoats are always alert and inquisitive and they often sit upright, showing, in the summer, the marked division between the creamy-white underside and the brown flanks and back. Their inquisitiveness has also brought about their death. Gamekeepers often used the trick of imitating the sound of a squealing rabbit to draw a hungry or curious stoat into the open, where they were then shot. However, it is also a useful trick for photographing them.

The usual gait of the stoat is a bounding one as it moves in leaps of nearly two feet. Stoats climb very well and will raid birds' nests for eggs and young and have been disturbed from squirrel dreys. They are also adept swimmers and there is one report of a stoat which was found in the water over one mile from land!

WEASEL

As with the stoat, the collective name for a group of weasels is a pack, though for most of the year males and females occupy different territories. The only time the collective name is appropriate is when a family party is playing or hunting. The weasel is Scotland's smallest carnivore. The head and body of the male is only 8 inches long and the female a mere 6½ inches. Their small size – even the male is only 4 ounces – is more than compensated by their fierceness in hunting. They are found throughout mainland Scotland but not in the Western Isles. They are most often seen when they cross a road and apart from their small size, identification is apparent by the lack of the stoat's black tip to the tail. Weasels also sit upright to get a better field of view but unlike the stoat the line between the whitish underparts and the brown flanks is irregular, with no two flank patterns the same. In Scotland they do not change their coat to white for the winter although this does happen further north in their range.

Weasels – sometimes called whittrets – occur in a range of habitats from farmland to woodland and hedgerow to stone dykes. They often move in un-dulating bounds about 12 inches between take-off and landing, although in contrast they can run very fast. The prey of the weasel is normally mice and voles taken on the ground but they climb readily and will get through a hole in a nest box – even if it is only 1⅛ inches in diameter – to get at eggs, chicks or adults. But their favourite food is mice and voles so they are friends of the farmer and forester. However, they are still persecuted by gamekeepers and large numbers are trapped each year. Like the stoat, prey is killed by a bite to the back of the neck.

The tiny young – called kittens – are born in a nest of leaves or grass in a hole or crevice. Unlike the stoat there is no delayed implantation of the fertilised egg and the young are born in April or May with an occasional second litter in July or August. The young stay with the female until they are 12 weeks old by which time they have reached maturity. Unlike other British carnivores who do not breed in their first year, the young weasels from an early litter may breed later that same year. Intriguingly the vole plague of 1892 in Scotland was blamed on the fact that large numbers of weasels had been live-trapped for exportation to New Zealand.

MINK

Escapes of non-native mammals into the wild have caused problems as typified by the rabbit and the sika deer, but perhaps the most calamitous one of all has been the North American mink. Mink were originally imported in the late 1920s with the aim of breeding them in captivity for their valuable pelts. But people under-estimated the ability of this carnivore to escape and since 1930 they have spread to most parts of Britain and are still spreading. Their full impact on other birds and mammals is unknown but on island locations they can devastate the eggs, young and adults of ground nesting birds.

Soon after mink were brought in to the 'fur-farms', there were reports of escapes and of feral mink living in the wild, but it was not until 1956 that the first breeding was noted. At first the animals were regarded as pests and they were trapped and shot whenever possible, and though thousands were killed it had little effect on the populations. There is a school of thought which says that we might just as well accept the mink as an addition to our wildlife as it seems

to be permanently established and in any case, eradication does not seem a possibility. It may well be that like the grey squirrel we have to allow it to colonise its own niche. Nevertheless there must be grave reservations as to the impact its diet will have on some of the rarer water birds such as grebes, divers and ducks and where it occurs in coastal situations, what its likely effect is on sea bird colonies. At present its effect on otters is unknown although it competes for the same type of food in freshwater habitats.

Mink dens have been found in tree trunks and in holes and crevices among stones including scree. The normal litter is five to six and the young are weaned after eight weeks, reach adult size in four months and can breed the following year. Original wild mink are dark chocolate-brown, which in the field looks black, and have a white chin patch and a bushy tail. The mink in captivity, however, were bred to a variety of shades and colours, but once breeding in a feral state the offspring soon revert to the original colour and pattern.

POLECAT

The extinction of the polecat in Scotland is still shrouded in mystery, although reference books give the final date as 1907. It has been suggested that a few pure animals were still hanging on in the north-west of the Highlands in the late 1960s, but their presence was never proven. Their extinction from Scotland was the result of intense persecution partly because they were classed as vermin and partly for their value in the fur trade. Reports of polecats in Scotland now always refer to polecat ferrets that have either been released or have escaped from rabbit trappers. The habitat formerly included woodland, farmland, marsh, river bank and plantations. The polecat was completely carnivorous and prey included hare, rabbit, small mammals and hedgehogs. Polecats mainly hunted on the ground as they were reluctant to climb or swim. It was known as the 'foul-mart' because of its strong smell while the pine marten was called the 'sweet-mart'. Apart from man it had few predators and in its present habitat in Wales many are unfortunately hit and killed by vehicles.

Now extinct in Scotland, the last recorded sighting may have been in the late 1960s, but this was never confirmed. Since this time, all sightings refer to escaped polecat ferrets.

Badgers and their setts are fully protected under the Wildlife and Countryside Act.

BADGER

Whilst it is quite possible to see a pine marten or wildcat in broad daylight – more often than not crossing a road – the same cannot be said about the badger which is far more nocturnal in its habits. In Scotland they live in woodlands, banks, sometimes in fields and even in block scree in parts of the Highlands. They spend the day in holes called setts and these large and sometimes extensive tunnels and chambers are dug out with their broad and very powerful claws. The badger is thinly distributed in Scotland and despite their size, the setts can easily be overlooked in sparsely populated areas such as the Highlands. The sow has two or three cubs, normally in February, which rarely come out of the setts before eight weeks. Although weaning starts at twelve weeks, some suckling may continue up until four months.

Watching badgers at their setts is not as difficult as it may first appear although the welfare of the animal must come first and they should not be unnecessarily disturbed. Scent, movement and noise are the keys to badger-watching success. Scent is probably the most critical as this is the badger's most well-developed sense, so you should hide downwind from the sett. Semi-camouflaged clothing is advised so that the observer blends into the background, but it is important to keep still and make no noise. Some binoculars gather in a great deal of light and can be used in strong moonlight, but a better idea is to visit the sett well before dark, especially in the summer months when the shorter nights mean a better chance of the badgers coming out to feed before it gets really dark.

If you have a strong torch, this can be used providing you cover the front with red cellophane or some other transparent red material, as badgers cannot see red light. Once you have found a sett – searching woods and hedge banks is one way and looking for tell-tale greyish hairs on fences is another – there is the problem of finding out whether a sett is occupied. One trick is to visit the sett in daylight and place some small twigs in the soil at the entrance to the sett, leaving about six inches of the twig sticking up. If you visit the sett the next day and find the twigs knocked over, then it is a sign that the sett is occupied. This is only a sign, however, as sometimes another badger may visit a sett and sniff at the entrance, knocking the sticks over before he continues on his way. Rabbits and foxes will also occupy badger setts that are not in use.

One of the basic foods of the badger is earthworms, which it will dig for or take off the surface of the ground after rain. In addition they have a broad taste and their diet can consist of cereals, beetles, fruit in the autumn and some mammals – particularly young rabbits dug out of their burrows. In the Highlands badgers probably rely more on carrion. A rare record from a glen near Inverness describes how at 1,500 feet and in the middle of the afternoon a badger was seen eating off a red deer hind carcase. Contrary to expectations badgers are often active in the depths of winter (the only land mammal to truly hibernate in Scotland is the hedgehog) although they do put on fat reserves, bringing their weight up to around 25 pounds in the autumn, so they do not necessarily need to find food every day in the winter. Tracks in fresh snow have revealed that a badger will easily travel two miles from its sett to search for food.

Unfortunately man has always been a predator of the badger, whether by catching them in gin traps in the old days, or by killing them with poisoned bait such as rabbits. We tend only to hear about the birds of prey that are killed in such ways but with the badger and the slow-acting poisons of today, it will usually go underground before it dies in agony. This and natural deaths underground mean that when setts are regularly cleaned out (the badger is very hygienic) the bones and skull are often thrown out near the sett.

Perhaps the most barbaric of all cruel persecutions is the practice of badger baiting that still exists in some parts. Dogs are put to the badger, which is deliberately injured in some way before the fight to stop the much more powerful badger from killing the dogs. In Hugh Miller's day he recounts that many inns had a badger pit under the floorboards in the bar and anyone who wanted to pit his dog against the captive badger could do so – with side bets on the outcome.

Otters are more abundant in the north-west of Scotland.

OTTER

In parts of Scotland you still stand a good chance of seeing an otter but it is more likely to happen on the western seaboard of the Highlands, the Western Isles or in the Northern Isles. If you are lucky the otter will find you, as has happened to many a fisherman on loch or sea who has suddenly had that uncanny feeling of being watched, only to find an otter inquisitively looking at him.

In the mid-1950s there was a decline in otters which was probably due to pollution by organo-chlorine insecticides. These also severely reduced the number of peregrine falcons, sparrow hawks and golden eagles but whilst these raptors made a comeback, the otter did not in many affected areas. This critical reduction in numbers means that today they are particularly vulnerable to the wide range of ways in which otters are still killed, both deliberately and accidentally. People kill otters because they believe they adversely affect fish stocks, although the arguments against this are strong. In any case, the current number of otters is so low now that it is extremely unlikely to have much effect at all on fish stocks. This is also true for fish farms, where some people kill both otters and seals. The recent upsurge in catching eels in fyke nets is worrying as some of the nets may not have been fitted with anti-otter/bird guards. One of the worst examples of otters being killed in such fyke nets was in the Western Isles where in 18 months of 1975-76, no less than 23 otters were found dead in eel fyke nets. The otter is now fully protected under the Wildlife and Countryside Act of 1981, but as with all legislation the implementation is another matter.

The ban on the use of the gin trap in Scotland has undoubtedly helped the otter considerably, but there is still evidence of predation by man, particularly by shooting. While there has been a decrease in the number of estate staff in recent years and thus perhaps less shooting of otters than in the past, with the increase of vehicles and traffic, many more otters are now killed on the roads. But despite all these predation factors, the otters' future may be decided by such aspects as silage effluent, fertiliser run-off, in-dustrialisation and acidification of freshwater that adversely affects fish stocks. Even on their own, these factors can cause fragmentation of otter populations, but collectively they spell disaster. Oil pollution from spills can also be a serious problem as the oil ruins the waterproofing of the fur and can result in the otter dying of exposure.

The movements of the otter are less predictable than those of many birds and animals. While it is well known that they are active in the hours of darkness, it is surprising how often they can be seen during daylight hours. Perhaps the best time to see otters is in the summer during the first two hours of daylight. Sheltered bays on the coast are probably the best places to look, as choppy water makes the otter difficult to see. Some people believe that in Scotland there are sea otters and freshwater otters but they are one and the same animal as otters will often leave the coast to forage inland for food, running up the narrowest of burns and swimming in lochs before moving back to their holts on the coastline. At one time otters were commonly found living inland but too much disturbance by fishermen and hill walkers may well have driven many of these to the coast.

The otter is surprisingly large, around a metre in length, about a third of which is the thick, strong tail. They are equally at home on land or in water and find their food from a range of habitats. Their diet consists mainly of fish and the slower species such as eels are favoured. But they are great opportunists and will take a variety of food, such as the case where an otter ate two black-throated divers' eggs in front of a photographer in a hide, while he was photographing the divers. Another came up under an adult black-throated diver on a Ross-shire loch as the diver was swimming, but in this case the otter was disturbed and the diver escaped, only to be found dead on the loch-side the next day. In a further incident an otter was seen raiding herring gulls' nests and eating the eggs and it has also been reported that an otter on Skye actually took a chicken that was grazing on a shoreline. Other food includes frogs, birds, small

Territory is marked by 'spraints', often left beside a burn.

mammals and waterside birds. They are active hunters and even forage for food in mud and under stones. In most cases the food is caught in the otter's mouth, with the front paws only used to help once the prey has been seized. Otters are sometimes seen swimming on their backs with a fish or crab held between their front paws, but prey is usually taken to the shore and devoured.

Although otters are often seen just by chance, there are a number of ways in which their presence is indicated. Perhaps the best of these is the occurrence of droppings, called 'spraints'. The spraints are dark and slimy with a strong fishy smell, and often contain fish scales and fragments of bone. They are used to mark territory and are generally by or on apparent landmarks such as a promontory on a bank, a fallen tree trunk or on a large stone under a bridge or culvert. If the spraints are made on a well-used grassy site, the grasses may become stained brown and will be conspicuous. Otter tracks are also easily identified and a sandy coastal or freshwater shore is worth looking at. Each track is just about 2½ inches long and shows five webbed toes and there may be a groove between the tracks where the otter has dragged its tail. Another sign is an otter's slide, which can be down grass, a muddy bank or in snow – the animals just love sliding down into the water. Otter holts should not be

sought out, in order to prevent any disturbance to the otter.

In both the coastal and freshwater habitats, the dog and bitch otters normally live separately, coming together only during the mating season which, unusually, can be at any time of the year. A bitch is on heat for about two weeks in every five and the two otters may find each other by scent or their distinctive whistling calls. Before mating they are often playful with many mock battles taking place. They normally have two or three cubs which are born blind and toothless and their fine fur coats are not waterproof. They crawl after two weeks but the eyes are not open until a month after they are born. Solid food is eaten after seven weeks although they are still suckling over two weeks later. It is over two months after birth before they take to the water. The family group breaks up after about a year when the bitch comes into heat again. The last ten years or so has seen a significant amount of research into the otter and its ecological requirements, although we are still a long way from fully understanding its needs. There are now population figures for a large part of Scotland that can be used to monitor future changes in numbers but ironically the least surveyed areas are those that support the most viable populations, namely the bulk of the Highlands and Islands.

Adult otters are usually solitary, coming together only for mating.

WILDCAT

Although formerly found throughout Britain, apart from Ireland, the wildcat is now confined to the Scottish Highlands. A striped body and ringed, bushy tail with a blunt tip are key identification points of a pure wildcat. Domestic or feral tabbies normally have blotches in with the stripes and the tail is less bushy and more pointed. The wildcat's tail is also much shorter than the tabby's at around only 40–45% of the head and body.

Wildcats are solitary and territorial. They mark their territories with urine and probably droppings, although unlike the domestic cat the latter are not buried. Although they are agile climbers wildcats spend most of their time on the ground. They are active from dawn to dusk and seek their prey more by stealth and surprise than speed. Prey is often stalked but the wildcat will also lie in wait and then pounce. Scent is important to the wildcat, but a keen sense of sight and hearing plays a vital role in its survival.

It no doubt came as a surprise to many people in Scotland, particularly in the Highlands, when the wildcat was given even further protection after the first five-year review of the Wildlife and Countryside Act 1981. When the Act first came out the wildcat had the same strange protection afforded to the pine marten in that authorised persons could shoot it but they could not trap it live to be released elsewhere. Now it is placed on Schedule 5 of the Act which means almost total protection and the surprise registered by some people reflects the contrasting attitudes towards this species in the past and the very real problem it faces for its future survival.

Wildcats have for some unexplained reason always been very high on the so-called vermin lists of estates and by the first decade of this century it was almost exterminated by relentless persecution. It seems likely that the reasoning was an attitude of mind rather than a solution to a problem. The gamekeeper's gibbet was at one time a form of merit and the number of vermin and other species killed was taken to indicate the level of efficiency of the keeper concerned.

Traditionally carnivores were classed as vermin and anyone killing them was thought to have done the public a service quite apart from the fact that many carnivores had a price on their heads. Another look at the list of vermin killed at the Glengarry estate between 1837 and 1840 is most telling as 198 wildcats were killed with 78 house cats. The importance of this figure is that this is not an isolated case but is typical of the estates of that period.

Fraser Darling in his classic book *West Highland Survey*, published in 1955, states that 'The wild cat, almost exterminated by 1914, has made a wonderful recovery and is now numerous and widespread in the mainland Highlands. Persecution is now unlikely to overtake the rate of increase.' 1914 was a critical year because it saw a decrease in persecution as the call of World War I lowered the number of estate staff. This continued throughout World War II and it seems unlikely now that keepering will ever reach its former level in Scotland. The Forestry Commission played its own part in the wildcat's recovery by its afforestation of large areas. The wildcat sought and found refuge in these new lands, where it did no harm to the trees and was therefore not persecuted. Indeed it may have been helpful by keeping the levels of mice and voles under control.

The survival of the wildcat since 1914 has not been without its problems. One of the basic prey species of the wildcat, the rabbit, was decimated by myxomatosis and for a few years, wildcats were almost exclusively restricted to the small populations that the disease failed to reach or those that recovered quickly. But with the general recovery of rabbit populations and with the increased habitat from afforestation by the Forestry Commission and private companies, for a time life appeared good for the wildcat.

Today, however there is a new threat to the wildcat which is perhaps more serious than any before. It now appears that hybridisation between wildcats and domestic cats is so widespread that there are at present only a few relatively small populations of true wildcats left in Scotland today. These isolated

Scottish wildcats are threatened by widespread hybridisation with feral domestic cats.

groups of pure stock must be protected at all costs from further persecution and have therefore been given full protection by legislation. Even throughout Europe the wildcat is a rare mammal and is therefore listed on Appendix 111 of the Council of Europe Convention. The current problem of hybridisation must also be seen against the historical background of poisoning, trapping and shooting that still continues, despite its illegality. To a large extent it is an attitude of mind and old attitudes die hard in the countryside. But for the Scottish wildcat to survive, people must be educated to appreciate and respect it and all persecution must stop immediately.

You have to admire wildcats – perhaps no other mammal in Scotland has such a bearing and they are generally larger than expected – one shot at Altnaharra, Sutherland, weighed 12 pounds, with a body length of over 22 inches and a tail some 12 inches long. Apart from the occasional brief sighting as a wildcat crosses the road it is one of the most difficult mammals to see in Scotland, let alone watch. The best chance is to visit suitable habitats, such as old woodland with nearby block scree slopes, at first light in the summer and just keep very still and watch. The den may be in scree, an old fox earth or even under a tree stump. During the mating season in March, the males can be very noisy indeed. Two to four kittens are born in mid-May and are reared by the female. The kittens emerge from the den at five weeks, are weaned at four months and independent at five months. Their diet is varied and includes birds and small mammals up to the size of a roe deer kid.

RED DEER

Red deer are native to Scotland and thrived in the old wildwood and the vast 'Forest of Caledon' where Scots pine dominated the slopes throughout most of the north of Scotland. The development of peat, felling of woodland and the degradation of the uplands by overgrazing and overburning, still taking place today, caused the woodland cover to fragment and red deer were forced to adapt to open moorland. Where they can, the deer will still choose the option of living in woodland, whether conifer or broadleaf and these woodland red deer are generally heavier and darker than their moorland counterparts.

It has been estimated that the Highlands can support 150,000 red deer but by 1991 there were well over 300,000 and in many areas the population was out of control. There are many factors causing this overpopulation, such as a decrease in traditional estate staff, low venison prices, too few hinds culled and a series of mild winters that reduced winter mortality to an all-time low. At one time it was suggested that legislation be introduced forcing estates to cull a required number of red deer. However, this would have meant leaving carcases on the hill as it would not be financially worthwhile bringing them in to the larder. The basic problem is too many deer on too little ground.

An aspect to the overall reduction of the deer population is the market price of venison offered by the game merchants. This price is linked to the mystery as to why so few people in Britain eat venison. Most venison is in fact exported and it is a sobering thought that up to 80% of red deer venison from Scotland goes to the continent.

There must be a better way of dealing with such a resource as venison which it has been said is 'A Meat For Great Men'. The problems within the venison market lead to a fluctuating demand and corresponding fluctuations in prices which estates receive from game dealers. Demand reached an all-time low at the time of the Chernobyl disaster when high levels of radiation stopped the export of venison to the continent. The price per pound brought the market to its knees and there was talk of drastic reduction in estate staff and not even bothering to cull any deer at all. The market has recovered, but the prices are still low and face difficult times ahead. What is needed is a high-profile selling package to encourage more people to eat venison in the home market.

There are many myths and mysteries about red deer and one of the most popular is perpetuated by the famous painting by Edwin Landseer entitled 'Monarch of the Glen'. It gives the idea that the stag is the Monarch above all, whereas in actual fact it is the hinds that rule the roost. This becomes clear when a group of mixed stags and hinds are disturbed. The stags will be the first ones up and away to the next rise, leaving behind an old hind to see when danger has passed. Another popular myth relates to the antlers. It is often said that each point or tip on an antler indicates the stag's age in years, but this is completely untrue. Stags will occasionally live up to 12 years old, but depending on conditions, usually pass their prime well before that.

For many people, one of the most fascinating aspects of the red deer is when the antlers are used as weapons in the traditional 'fights to the death' which some consider as characteristic of the red deer rut in October. Whilst such dramatic fights do happen, they are far less common than most books make out, as the roaring or bellowing of stags is usually sufficient to settle differences before coming to blows. Antlers are shed, sometimes called 'casting', in the spring and re-grow for the autumn rut. The growing antlers are protected by a soft skin called 'velvet' and are very sensitive. If any fights break out during this growth period, the stags will not risk damaging the antlers, and so rise up on their hind legs and box in much the same way as brown hares. The precise role of antlers is not completely understood as their annual growth can considerably weaken the stag. Those stags without antlers, called hummels, are generally stronger and heavier, important attributes a stag needs to hold a harem of hinds in the rut. Perhaps antlers were primarily used for defence in the days when the wolf was a natural predator of the red deer in Scotland.

Perhaps the epitome of the wild Highlands is the period of the rut in October when straths, glens and hillsides echo to the sound of red deer stags roaring at each other. The stags will wallow in peat hags and it is uncertain whether this is to cool them down or make them look fiercer and more dominant with peat hanging off the antlers. Challenges, whether vocal or physical, to the current dominant stag are frequent, with the respective stags eyeing each other up and strutting broadside to each other. During these challenges the hinds graze nearby completely unconcerned, almost as if the whole affair had nothing to do with them.

For most of the year hinds and stags live apart, although young stags will often stay with the hinds for a couple of years. The calves are born in the spring and are occasionally taken by fox and golden eagle, but this is rarer than some books suggest. However large numbers of deer die each year and provide the some 400 plus pairs of golden eagles in Scotland, who take about 70% of their food as carrion, with a convenient source of food. Unlike the roe deer, twins in red deer are very rare. Calves are often found with no hind in sight, but it is unlikely that they have been deserted and should be left where they were found. The hind will simply have left the calf and wandered off to feed and she will soon return to feed her offspring.

With the overpopulation of red deer completely out of control under the existing system, the general condition of the deer is poor in many places. There is increasing pressure by deer on agricultural land and in particular the increasing areas of afforestation. Red deer are now seen where there have been no red deer in living memory. Another problem of the present overpopulation is that a really severe winter will kill tens of thousands of red deer and also weaken the condition of the remaining herds, which is a waste of a valuable resource. The current management of culls places too great an emphasis on stags – hunters will simply pay high prices to shoot stags for the trophy of the antlers. Under this pattern of culls it seems that numbers of red deer will rise even further. Indeed, perhaps the only way forward is to introduce legislation that will force landowners to cull a sufficient number of both stags and more importantly, hinds. The problem of such legislation would be its enforcement, although the annual counts over large tracts of Scotland by the Red Deer Commission would reveal if the landowners were acting responsibly and their culls were adequate.

SIKA DEER

Although the introduction of the fallow deer is poorly documented, the introduction of sika deer is well recorded and began as recently as 1870. Sika deer were introduced to the fashionable deer parks that held the usual favourites of red and fallow deer. However, for the long period of the two World Wars, fences were neglected and many deer of all three species escaped. Even so, the current population has not arisen totally from 'escapes' as many were deliberately introduced into the wild for sporting purposes. These sika deer did surprisingly well in the harsh climate of the uplands.

The beginning of the 1970s saw an apparent increase in the numbers and distribution of sika deer – possibly because of a series of mild winters. In the middle of the 1980s, a small number of hybrids between sika deer and red deer were reported. In one such incident near Lochinver, Sutherland, a small number of sika stags appear to have crossed several miles from the east over open moorland to feed in the birch woodlands along the west coast. Late in the year when the rut started, which is at the same time as the red deer's, the sika stags found themselves without any sika hinds and so some hybridisation took place with the red deer hinds. This sent a minor panic through the deer world with some seeing it as the end of the pure strain of red deer in mainland Scotland. The only true red deer of the future would be found on marine islands to which the sika deer could not swim. This, however, has not happened and hybridisation remains a rare event.

Sika are mainly woodland animals, preferring the dense cover of birch and hazel thickets or brambles and bracken. They are, in fact, very difficult to see in these habitats as they have a practice of standing motionless until any danger has passed. In daylight, sika remain hidden in the woodlands but during the hours of darkness, they leave the cover to browse and graze in glades, rides or open ground. This pattern of behaviour means that unlike red deer, they are very difficult to count and therefore, while their distribution is known, there has been no attempt at accurate census counts. The Red Deer Commission have based their estimate of 9,000 to 11,000 sika deer in Scotland on the cull returns from estates.

Stags and hinds live in separate groups in February and March and calves are born in June or July. Stags may remain solitary until the rut in October and November at which time they wallow in peat hags in the same manner as red deer stags. During the rut one of the stag call notes is a penetrating whistle which has a most eerie effect when it comes from dense cover in which the stag cannot be seen. Like red deer the sika deer shed their antlers and regrow them for the rut. Sika deer will graze on grassland and rough herbage but have a fondness for the shoots of hazel. Bark is stripped from medium-sized branches as compared with the trunks of small trees so favoured by red and roe deer.

Apart from the loud penetrating whistle, sika have various vocal and visual ways of communicating, though we do not know the meaning of all of them. For example, hinds with calves will make a soft whining note for no apparent reason, although it may be simply a contact note. When sika deer are alarmed their stiff-legged gait and extension of the rump patch hairs, called 'flaring', are highly pronounced. The area of the heart-shaped rump patch can double when the flaring takes place and is that much more conspicuous than even the red deer's because the patch has white hairs and a white tail. During such alarm behaviour the deer call with a strange sound that has been likened to a cross between a squeal and a sharp whistle.

As with red deer their main predator is man. However their calves are smaller than red deer calves and predation by foxes may be more important than is generally realised. Golden eagles are unlikely to be a threat because of the deers' preference for dense woodland. As for the future of sika deer it would seem to be secure, especially with the continuing mild winters and more and more areas of afforestation. It could well be that this species will spread into every suitable area in Scotland and only a series of severe winters could alter this dramatic and successful colonisation of an introduced species.

FALLOW DEER

Fallow deer are mainly woodland deer and can be found in conifer or broadleaf woodland or a mixture as long as there is sufficient cover. As with the other woodland deer – sika and roe – they will, however, come out of the trees to graze and browse in rides or glades and they will often use well-worn tracks to reach fields for feeding on grass or root crops. Their distribution in Scotland largely corresponds with the former distribution of deer parks in which fallow were particularly welcome, as they were relatively tame, mild mannered and attractive, occurring in a variety of colour forms, including white. They are also generally easier to contain by fencing than the heavier red deer and much smaller roe deer.

Unlike the native red and roe deer the origin and former distribution of fallow deer is intriguing. While reference books normally put them down as a species introduced by the Normans in the Middle Ages, there is another interesting theory. This suggests that the fallow deer occurred in the last interglacial period, but died out leaving no trace of their presence. It is certain that remains of other deer species have been found at Roman or earlier archaeological sites, but there have been no confirmed remains of fallow deer. Yet another theory is that fallow deer were brought in by the Phoenicians. But whatever their source or origin, in modern times they were introduced for hunting, venison and to stock deer parks.

There are two colour forms in summer with arguably the most attractive having a rich fawn coat with many prominent white spots on the back and flanks. The other form is nearly all black, which is the form the Normans were reputed to have brought in and is dominant in some park herds. However, there are many other colour varieties, including white, and some parks have bred complete herds of this white fallow deer. Such is the case at Berriedale on the east coast of Caithness, where in about 1900 the Duke of Portland introduced some white fallow deer which he brought up from Welbeck Park in Nottinghamshire. By 1950 the herd had been reduced to about ten beasts and there are about the same number today. They are easily seen through binoculars just to the north of the main A9 as it starts to dip sharply into the village of Berriedale. This is now possibly the most northerly herd of fallow deer in Scotland.

The first three months of the year see the bucks and does living apart in herds. Then the herds of both sexes break up with the bucks casting their antlers in April and May and does giving birth to their young, called 'fawns' in fallow deer, in late June and July. As with the other species of deer, the young are frequently left alone by the hinds who have simply gone off to feed, and should not be mistaken as deserted! The fawns are born singly – twins are very rare in fallow deer – and their typical colour is chestnut-brown with white spots. Another form is blackish with brown dappling and the rare unspotted sandy fawn will grow up to be a white deer. The antlers are fully grown and harden off for the rut in October when the woodlands echo to the noise of the rutting bucks who make a deep and loud belching noise called 'groaning'. The herds re-form for the rest of the year although the older bucks will stay on their own.

The rut is the most exciting time to watch fallow deer as the noisy bucks gather their harems around them. During this time the buck rubs his head which contains scent glands against saplings to mark his territory and this often frays the bark and brings him into conflict with the forester. He also thrashes his antlers against branches and bushes and will occasionally use a wallow in the same way as red deer stags. Most confrontations between bucks are resolved by simply making a great deal of noise. But when fights do occur, they can be fierce with clashing of antlers and furious charges until one buck retires hurt or defeated and the winner takes the harem over until

the next contender appears. The antlers of the bucks are easy to distinguish from red or sika deer as once the buck is four years old each antler has a flattened broad palmate area. Stalking fallow deer has never been quite the same as the traditional stalking of red deer, perhaps because fallow deer still have a 'park' image about them. However as woodland deer, they can be difficult to stalk, though they have one type of behaviour which can often be their downfall. They will stand motionless when a person goes past a thicket just the same as sika deer, but unlike sika they will then step out of cover to watch the person go away – often a fatal mistake if the person is a stalker.

Fallow deer will browse on the new leaves of many deciduous trees and in some areas bark stripping is a problem for the forester. They can also affect conservation efforts when an area of scrub is cleared and trees such as willows and hawthorns are left to provide valuable habitats for insects. Fallow deer just love willow bark and any trees isolated by clearing will normally be de-barked within 24 hours. But they also take a wide range of fruits and berries – from acorns to rowan berries and blackberries to hips.

The bark of felled conifers is an important winter food. For some unknown reason fallow deer rarely eat certain plants that are sometimes abundant and available all the year round such as purple moor grass, tufted hair grass, bracken and gorse. Feeding times are largely affected by disturbance and where this does not happen fallow deer will feed at any time of the day or night. Where disturbance such as shooting occurs on a regular basis, the deer can be very secretive and will only feed during the hours of darkness.

ROE DEER

Roe deer are by far the smallest of the four species of deer in Scotland with the buck measuring only 25 inches high at the top of the shoulder. As a native species it was at home in the old forests and was widespread in the Middle Ages. Unfortunately being mainly a woodland deer, the vast clearances of forests have led to a drastic reduction in the number of roe deer. Unlike the more adaptable red deer that simply moved to open moorland, the roe deer could not cope with the open exposed conditions. This, however, is not likely to be the complete answer to the mystery as to why, after the extensive felling of woodland in the 18th and 19th centuries, the roe deer was almost extinct in all the Scottish counties south of the Edinburgh to Glasgow road and only local to the north of this line.

The reason why they are now widespread throughout Scotland is a combination of factors, but despite a few deliberate introductions, the spread has mainly been a natural one both in terms of distribution and numbers. Afforestation of all types of woodland has undoubtedly helped as new plantations are ideal for roe deer and they have also changed their habits in recent years and can increasingly be found on open ground. A good example of this is the resident population that now exists on open moorland near the top of the Drumochter Pass, just north of Perth. The present series of mild winters has also assisted in the increase in numbers as winter mortality has been so low.

Part of the problem over the uneasy alliance between the roe deer and the forester is that no deer fence has been designed that will keep roe deer out, as they can get through very small gaps. If a stalker shoots a red deer in a forest he knows that unless there is a hole in the fence, another red deer is unlikely to get in. With roe deer it is a different story and for a long time it was thought that a buck in his territory was worth leaving alone as he would stop any others coming into the wood. Nowadays every effort is made to take roe deer out as soon as they can be culled. As with all species of deer in Scotland, if roe deer are in enclosed areas and likely to cause economic damage to trees, they can be shot whether in or out of season. One problem with roe deer is that they regularly browse on trees on the edge of rides or glades and it is so easy to jump to the conclusion that this limited damage is widespread throughout the wood.

Red deer regularly swim to freshwater islands to reach better grazing but there are few records of roe deer following their example. However, a notable exception happened near Lochinver in Sutherland and not in freshwater but in, of all things, the sea. In the 1970s, single roe deer regularly swam three quarters of a mile in the sea to feed on the comparatively lush grazing on a marine island. After a few years a small number stayed all year round and although the island is only 15 acres in size a small breeding group continues to flourish.

The seasonal behaviour of roe deer is more complex than the red deer's. They stay in small groups, never more than a few individuals, although on good grazing they can appear to be in slightly larger numbers as the deer move out of woodlands to feed. The buck casts his small antlers in November or December which are extremely difficult to find on the woodland floor. With red deer the antlers regrow for the rut but the roe bucks' antlers are fully formed by late spring whilst the roe rut is not until late July or the beginning of August. Roe antlers are generally around 10 inches long and rarely have more than six 'points'. The antlers are sharp and despite the size of the stag they can be dangerous. This, together with the difficulty of fencing them in, has made them unpopular in deer parks. The doe has her kids – often twins and sometimes triplets – around May and the does are unusual in that there is a delay in the implan-

tation of the fertilised egg.

A roe buck in his sleek summer coat, which is foxy-red with a buff rump patch and black nose, is arguably the most attractive of the deer in Scotland. Like the sika deer the extension of the rump hairs is pronounced in the defensive mechanism of flaring. Despite their size roe bucks have been known to attack people for little apparent reason, so as with the other deer species, they should always be treated with respect. Roe deer are mainly browsing animals although they will graze in rides, glades and woodland margins. Feeding tends to be for short periods and in summer, broadleaf trees such as ash, hazel, and oak are important. In winter the main food includes heather, spruce, bilberry, hazel twigs and grasses. The feeding behaviour depends on a wide variety of factors such as disturbance, availability of food, cover and weather. The venison of roe deer is not particularly sought after in this country but on the continent it is considered a delicacy among the different types of venison and fetches a higher price from game merchants.

REINDEER

There is a long list of mammals that have been lost from Scotland's fauna, from the very early ones such as the mammoth and woolly rhinoceros, to the ones that have disappeared in historic times such as the wild boar, or the last large native mammal to become extinct – the wolf. However, perhaps the most exciting of the ancient animals would have been the numerous herds of reindeer that were common around 200,000 years ago. They must have been a remarkable sight as they moved between their summer and winter grounds. They occurred during the last glacial period but no one knows when they died out and there have been no positive fossil finds to suggest they survived into the post-glacial period. The remains of reindeer from Iron Age sites in Scotland are probably the result of imports from Scandinavia. Whilst there is a reference to reindeer in records of the 12th century Orkneyinga Saga, it seems more likely that these were red deer, although again reindeer from Scandinavia might have been seen.

A domesticated Swedish herd of reindeer were brought in from Lapland in 1952 and re-introduced in the Cairngorm mountains near Aviemore and have been managed there ever since. The principal feature of the reindeer is the huge antlers of the bulls. The species is unusual in the deer world as both males and females have antlers, although the cows' antlers are much smaller. In North America they are called 'caribou', which is American Indian for 'shoveller', after the way the deer use both feet and antlers to scrape away snow to find food. Their very large cloven hooves can be splayed out when they walk to stop them sinking too far into snow and the hooves click as they walk, making a noise that is characteristic of large herds of reindeer.

The social system in the September to October rut is similar to that of red deer in that a rutting bull will challenge rivals before mating. The calves are born in May or June, are unusual in not having spots and can walk within an hour of birth. Whilst the bulls cast their antlers in the autumn the cows retain theirs until the spring, so that in the winter they can use both antlers and feet to clear snow away for both their calves and themselves.

WILD GOATS

It does not really matter whether we refer to goat tribes in Scotland as wild or feral, as one thing is certain – they have been with us a very long time. Stone-Age farmers may well have brought goats into Britain and there have been tribes of goats living here for well over 1,000 years, wandering and breeding on inland cliff, sea cliff and open stony ground, where they thrive unless persecuted by man.

Kenneth Whitehead in his book *The Wild Goats of Great Britain and Ireland*, published in 1972, has a gazetteer of wild goats found in Scotland and lists no less than 141 sites. These 'tribes', the name for a group of goats, range from northern Sutherland down to the Borders. Argyll is at the top of these lists with a population of some 38 tribes, while at the bottom of the list, areas such as Nairnshire and Angus support only one tribe each. These lists are the only current source even trying to estimate the Scottish population and some of these tribes have indeed become extinct since the book was written. But if we estimate that the average tribe could be approximately 30 individuals – some are much smaller whilst others number over 100 – then the population could be over 4,000.

Tribes of goats have become extinct in Scotland for a number of reasons, but this usually occurs when they have come into conflict with man. The main reason for shooting out a population is either because of forestry interests – goats are notorious for eating trees – or because they supposedly compete with domestic grazing stock such as cattle and sheep. Ironically, tribes were introduced onto some estates because of their more aggressive and sure-footed nature, as they could graze cliff ledges while also keeping the sheep away, which otherwise have a habit of falling off such places. In the last two decades the introduction of a new tribe has been a rare event and I only know of one, near Inverness, where they were introduced for their sporting value.

However, the origin of most of the tribes is unknown, although there are probably a variety of reasons apart from introductions to save sheep. Originally goats were brought into Scotland during the Neolithic period when Britain was still part of the Continental landmass. In more recent times they were brought into Britain from Northern Ireland in the 19th century, but it is difficult to know just how many. It is possible that as they moved north they followed the old drove roads used for herding cattle to southern markets. A popular theory on the distribution of goats is that the farther north they were driven the more difficult it was to control them and escapes increased.

Increasingly wild tribes have been 'contaminated' by domestic goats which are deliberately or accidently released into the wild. These animals and their breeding influence on the tribe are easily recognisable as the domestic goats are often hornless and have tassels hanging from their lower jaw. They are also much larger and far more tame than their wild relatives. However, in a few years these characteristics disappear with successive breeding within the wild tribes.

Wild goats occur in a variety of colours with various combinations of black and white often dominating, but they also occur with brown or silver hair. The splendid horns of the billies have annual growth ridges on them; they are not shed like deer antlers and counting these ridges gives a good idea of the animal's age. It is a feature which can be identified with the aid of binoculars. The goats mate in October and November when there is a great deal of fighting between the billies, and when they raise themselves up and crash down, butting each other, it is dramatic in both sight and sound. The only damage is the occasional broken horn. The kids are usually born in January, often the most inclement time of the year, although some tribes living above 1,000 feet, will not have their kids until March.

Goats have few predators apart from man, although occasionally a fox or golden eagle will take a young kid if the nannies are not watchful. A few years ago there was a great controversy as to whether paying guests should be allowed to stalk goats. In many ways the idea is similar to the stag cull, with the trophy being the billy's horns, which can be very impressive indeed. Estates argued that goat numbers had to be contained because of the effect on grazing and on trees and shrubs. Naturalists argued that whilst accepting that a cull was necessary in some circumstances, it should not be done for sport. However, if a goat is to be shot, the important thing to remember is that it does not matter so much who pulls the trigger or if that person pays to do so, as whether it is done humanely. Such income can help estates in the same way as the red deer cull.

Goats vary considerably in their tameness with some tribes, like the small, dark goats south-west of Fort William, being very confiding. This tribe occupies a cliff on the edge of a sea loch and they will graze quite contentedly at the roadside although they move off slowly if a car door is opened. In contrast if a tribe is regularly culled they can be very difficult to approach on the open hill or cliff and they can move surprisingly fast. In such instances they can be as difficult to stalk as red deer. Some tribes are hefted to certain areas but it is known that there is occasionally some interchange between tribes even when they are a few miles apart.

Until the coming of sheep to the north of Scotland and when the cattle were annually driven south for sale in late summer, the goat was very much prized as it could survive the winter without any supplementary feed. For a long time it was known as the 'poor man's cow' and it was a valuable source of hair, meat and milk. Goat's milk would be far more popular today if it were not for the prejudices established against it in the early part of this century, when it was supposed to have caused brucellosis in Britain, whereas the truth of the matter is that cow's milk was the culprit.

Goats have figured extensively in folklore and legend and became an integral part of religion and witchcraft, but there have also been more practical uses for them. Some people used to keep goats with horses as they were reputed to calm them down. If they were hunting horses, the presence of a goat meant that they would never kick out at the hounds. Likewise if a horse refused to board a ship, a goat was led ahead and the horse would follow. This gave rise to the expression 'Judas Goat' – a goat which led stock into an abattoir for slaughter. There is also the in-

The age of goats, especially billies, can be told by counting the annual growth rings on their horns. Their rut is in October and November, and the kids are born early in the year.

teresting theory that goats are reputed to keep the snakes away from an area. A small number of goat tribes owe their origin to this belief. Such is the case of the farmer at Torrachilty, near Inverness, who introduced goats in about 1880 to keep down adders.

But today there may be other changes in the world of the wild goat as increasing interest is being shown in them for their hair. Currently a wild nanny can fetch a good price in the market only as a source of hair. Obviously goats used for hair could not be allowed to roam in the wild but would have to be penned for easy access, so it may well be that domestication is the future for some of our tribes. With the combination of sporting interests, demand for hair, culls because of forestry interests and changes in grazing demands, the future of our wild goats must be treated with caution. It would be a sad day if the wild goats were to become extinct in many more areas.

SOAY SHEEP

The fact that Soay sheep are the only ancient sheep of Britain that are living in a truly wild state and that they are only found on the islands of Soay and Hirta in the archipelago of St Kilda, is but one fascinating aspect of these remote islands. Ironically there is no island called St Kilda, as the name refers to the archipelago, and no saint called Kilda. Not only do these remote islands – they are 100 miles west of the Scottish mainland – support these unique Soay sheep but they have also been isolated long enough to have their own subspecies of wren and woodmouse. This same isolation has enabled the Soay sheep to thrive for centuries. Their origin is obscure, but their fore-bearers were probably brought to Britain by Neolithic farmers. Another theory is that they were placed on the island of Soay by Vikings so that each year when the Vikings went raiding down the western seaboard they could call in for fresh meat. However,

the name Soay is Norse for sheep, so they knew it as sheep island and the sheep may well have already been there when they first landed.

Of the three main islands, Hirta, Soay and Dun, only Soay has a true untouched population of Soay sheep. When the St Kildans left Hirta in 1930 they took all the sheep that were on Hirta with them, although it is reputed that a few escaped capture. Two years after the departure of the St Kildans, the Marquis of Bute transferred a flock of sheep which included 20 tups, 44 ewes and 23 lambs from Soay to Hirta, where they have since thrived. Very little is known about the sheep on Soay as access is difficult, but the sheep on Hirta have been studied intensively in the last few years. Sheep on both Soay and Hirta are not managed in any way but their numbers fluctuate enormously in cycles of peaks and troughs. Annual counts on Hirta indicate that the lowest numbers are just over 600 but in contrast they peak in some years to over 1700.

The reason for this dramatic fluctuation is not known at present, although it is probably a combination of food shortage when the numbers peak and parasites. Current research may well mean that we shall soon know the causes and if it is food and parasites both of these can be controlled by supplementary feeding and treatment. This would then pose conservationists, particularly Scottish Natural Heritage who manage the National Nature Reserve, with a wildlife dilemma. If it were possible to stabilise the population on Hirta, or for that matter Soay, should anyone be allowed to start 'managing' such sheep when they have survived so long without the interference of man? On the other hand what would happen if the cycle of numbers was broken because of freak winter weather and instead of just a trough, the whole population died out?

The sheep have two colour phases, dark brown and fawn and both of these resemble the wild mouflon but are smaller in size. They have the same colour pattern of the mouflon but with lighter hairs around the eyes, chin and on the legs, with a white belly. The heavy, dark and curving horns of the rams are also similar in shape to those of the mouflon. As with wild goats the horns are not cast annually but growth is added each year. This annual growth leaves a series of dark rings around the antler and as with goats, indicates the age of the animal. The horns of the ewes are much smaller and sometimes absent. Soay sheep are timid but inquisitive on most of Hirta and all of Soay. However, in the old village on Hirta they have become used to visitors and can be approached within a few yards.

At first sight, the sheep on Hirta may appear randomly scattered but in fact they generally live in small groups and their social behaviour is rather similar to that of red deer. The ewes and rams live in separate home ranges until the rut in October when the ewes may stay in the same flock, but the rams start to separate and wander. As with the red deer, the stronger and more dominant males hold a harem together until they are superseded by another male. In this way the bulk of the lambs are sired by the strongest males. During the rut the rams eat less and less as they chase more and more ewes or try to break into a harem group, defend their own from other rams or stop their own ewes wandering. This lowering of condition if followed by a bad winter, can lead to some mortality by the time spring comes around again.

Winters can be severe on St Kilda with strong winds and prolonged rain and with the very short days and long hours of darkness, the Soay sheep seek sanctuary in the hundreds of turf-capped stone structures called 'cleitean'. The St Kildans built these fascinating structures to store many things including birds, meat and feathers. With the turf roofs keeping the rain out and the loose stones allowing a drying wind to pass through, they make ideal retreats for the Soay sheep. In the winter they spend the night in the cleitean and emerge only to feed on nearby grassland. In the summer there is a different pattern of movement. The sheep will still find a cleit or wall high up on a slope at night, but at daybreak they re-form their social groups and move down to the lower grassland to graze. The rams are the last to come down in the morning and the first to go back up in the evening.

Many Soay sheep have over the years been taken off St Kilda to form flocks on mainland Britain and there are now sufficient numbers in captivity for them to be self-perpetuating without further animals being sent off St Kilda. They are often kept in parks and referred to as one of the 'rare breeds', and they have on occasions been used in quite a specialised way. Their light weight means that they can graze quite sensitive vegetation without causing any undue erosion to the vegetation or soil. One problem with them, however, is that they cannot be rounded up by a sheep dog in the usual way as they simply scatter. The St Kildans with the help of their dogs would run the individual sheep until it was exhausted or cornered against a wall. Despite the fact that Soay sheep are an attractive sight in captivity, they are undoubtedly at their best on the wind and sea-sprayed grassy slopes of Hirta and Soay on the archipelago of St Kilda – 'The Islands on the edge of the World'. It is essential that the future is assured for this most primitive domestic form of sheep in Europe.

RABBIT

The rabbit is one of the many mammals to have been introduced into Britain. Scotland is no exception where it was deliberately released in many parishes. They were first introduced into Britain in the late 12th century and were originally kept in enclosed warrens and it is not clear when they became feral on a widespread scale. Originally, the rabbit's value as meat and fur was considerable, but before long it became clear that this value was out of proportion to the damage it did to agriculture. The amount of damage to crops today is less dramatic in comparison to what the former rabbit populations could do before the introduction of myxomatosis from France in 1953. As much as 50% of crops could be eaten by the rabbits and on evenings and early mornings the fields seemed to be alive with rabbits. Not only would they eat grass and crops but they also de-barked trees in winter and when they occurred in large numbers they would kill off young trees. Myxomatosis killed 95% of the rabbits in Britain and brought about a large reduction in the numbers of buzzards, foxes and stoats. Today their numbers are increasing again as the rabbits become more resistant to the disease which still affects them every year. In a few areas the rabbit population is back to pre-myxomatosis days and is causing considerable damage to crops once again.

The rabbit was originally called a 'coney', with the name 'rabbit' used only for its young. They are preyed upon by almost every meat-eating mammal in Scotland and all the larger birds of prey – from buzzard to golden eagle. The survival of the wildcat in Scotland is reputed to have been because there were small pockets of rabbits that were isolated and did not catch myxomatosis. The wildcats managed to survive on these rabbits until other populations increased again. In contrast, the efforts of man to eradicate or even control rabbits have gone to unbelievable lengths and costs, but in general it has not made much overall impression despite the numbers killed. Ironically, the rabbit was completely absent from the northern mainland of Scotland at the beginning of the 19th century, although abundant 100 years later. This corresponded with the drastic reduction of predatory birds and mammals in the reputed interest of game preservation.

The reason for the rabbit's success against all odds is partly due to its ability to flourish in almost any area with grazing and room for burrows. If there is enough cover, rabbits will live above ground all year around and such populations are not supposed to have contracted myxomatosis because the flea carrying the disease is normally contracted in burrows. The habitats can range from the shelter of woodlands to wind and salt spray on islands such as the Monach Isles in the Western Isles. The doe will rear as many as 20 young each year and some of these will breed when only four months old. Mortality, however, is high with rabbits dying from cold, predators, wet and diseases and it is likely that many live for less than a year.

There are very few problems with the identification of rabbits although young brown hares are similar, but have black tips to their ears which are absent in rabbits. Black rabbits are not uncommon and are sometimes called 'poachers' rabbits'. This name is derived from the practice of keepers who placed down a black rabbit in order to detect poachers, who would not notice their colour at night. If the black rabbit disappeared, the gamekeeper would know that it was likely that poachers had been at work.

Other colour forms occasionally occur and often come from tame rabbits that have been released or escaped, as they will readily interbreed with the wild ones. There is a mounted yellow rabbit in a case in the Inverness Museum which was taken at Coignafearn,

just south of Inverness, in 1928. This sandy form is reputed to be a throwback to the original enclosed warren rabbits that were mainly of this colour.

A colony of rabbits may look like chaos with rabbits running around anywhere, but this is not the case above or below the ground. There are dominant males and females who occupy the best parts of the warrens and therefore breed more successfully. Territory is marked out by what is called 'chinning', which is when a rabbit rubs its chin containing scent glands on the ground to mark out its boundaries. Small mounds such as ant hills are used by rabbits as lookout points and droppings on these will also be a sign of territory. Alarm is shown to others by a rabbit thumping its hind foot on the ground. When running, the white underside of the tail, the 'scut', also serves as a danger signal. Heavy grazing immediately around a rabbit warren leaves short vegetation so that rabbits can more easily spot any predator and their prominent eyes are placed so that they have a wide angle of view. Most of the day is spent underground where they eat the soft droppings that have passed through their system, which still have some proteins and vitamins in them. The second much harder and darker dropping are made above ground.

BROWN HARE

Brown hares are scattered throughout Scotland wherever there is farmland or rough grazing. They are also found in moorland, but they are absent from the north-west, which may be related to the heavier rainfall compared to the east and south. They are often seen running along a narrow road in front of vehicles, behaving as if they do not know what to do just before darting off into a field. They are also frequently seen during the display season in March when they become the famous 'mad March hares'. During this time two or sometimes three adults will get up on their hind legs and box with their forelegs, though they never seem to come to any harm. The boxing may well include a female making a point to an over-amorous male. It is an intriguing part of the aura of mystery that surrounds the brown hare, that the otherwise solitary hares will assemble for these spring gatherings, sometimes travelling some distance to a traditional place for display. During the boxing season they seem to ignore almost everything else happening around them and often fall prey to predators at this time.

In the wild, brown hares can live in quite exposed situations such as open fields but, unlike the rabbit or the mountain hare, they do not resort to any sort of hole. They simply make a depression in the ground in long grass called a 'form'. Despite being quite large animals, much larger than rabbits, once lying low in a form with their long ears flattened along their backs, they remain motionless and are very well camouflaged. They can be distinguished from rabbits by their longer black-tipped ears and loping gait.

Brown hares spend most of the day in or near their form, moving out to feed at night mainly on tender grass shoots or cereals. They are usually solitary but will occasionally gather in loose groups when feeding. Breeding is possible throughout a very long season – February to September – and the female can have three or four litters a year, each of two to four young. The young are remarkable in that they are born complete with fur and their open eyes. The 'jill' (the male is called a 'jack') leaves the leverets in a form close to their birthplace and once a day for the first four weeks of their lives, the leverets gather at sunset to be fed by the female. This is the only parental care they receive. As is often the case with mammals, the mortality rate is high and few brown hares make it past their first 12 months.

There is no legislation protecting the brown hare and opinions on their future vary, although one thing certain is that they have declined in Scotland this century. The reasons for this decline are complex but it is mainly because of changes in agriculture and ironically, in the changing fortunes of the rabbit. When the numbers of rabbits were decimated by myxomatosis, which never affected hares, the resulting increase in rough grazing provided cover for the brown hares and their populations increased. However with the more recent intensification of agricultural practices, the application of pesticides and the use of heavy machinery, there has been a dramatic loss of rough grazing, providing little shelter or food for the brown hare and it has consequently declined. In an ideal situation the brown hare needs patchworks of small fields allowing a diversity of cover and food. Intensely farmed areas, where there are large fields of a single weedless crop, are useless to the hare, and they have been known to starve to death between the harvest and the spring crops when the stubble has been burnt and ploughed.

Shoots are organised in some areas to kill hares and hare coursing using lurchers or greyhounds is still commonplace. In the wild, natural predators depend on whether the animal is an adult or a leveret. Adult brown hares are quite a handful and can take some catching, as they can run as fast as 35 miles an hour, although some are still taken by foxes, golden eagle and possibly wildcat. Leverets are an easier prey and are taken by buzzard and stoats. The brown hare is normally silent but will scream when frightened or in pain. Hares will often take to the water and swim well. The brown hare is probably the most widely distributed mammal in the world today.

MOUNTAIN HARE

Long narrow trails of depressed heather on tracts of moorland in the uplands are likely to be the tracks from the regular passage of mountain hares. The main area they are found in is heather moorland but this includes the arctic/alpine zone. As this specialised zone becomes lower in altitude in the north, it means that in parts of Caithness and Sutherland the mountain hare can be found almost at sea level. The mountain hare is indigenous to the Highlands of Scotland and Ireland and the Irish hares are not only a distinct sub-species but are characterised by the russet colour of their fur and the absence of the white winter coat. There are some yellowish mountain hares found in the Highlands and intriguingly some completely black ones in south-east Caithness, where 15 were shot at the beginning of this century and some of these are now found mounted in the Royal Scottish Museum in Edinburgh.

The distribution in Scotland of the mountain hare, sometimes called the 'blue' hare, 'verying' hare or 'arctic' hare, is confusing because of the many introductions to areas mainly for sport. The main introductions were in the 19th century and apart from the Scottish Lowlands they were also introduced to islands such as Shetland, Orkney, Outer Hebrides, Mull and Jura. A good example of these introductions is recorded in the game books of the famous Kinloch castle on the island of Rum where there are several entries indicating the introduction of hares for sport, which would have been mountain hares as opposed to brown hares. A few years ago the Nature Conservancy Council weighed up the pros and cons of bringing the mountain hare back to Rum, although any evidence of it being a native species to the island is flimsy, to say the least. The aim of such an introduction, or reintroduction as the case may be, was not from a sporting point of view but as prey for golden eagles and, possibly, the re-introduced sea eagle.

The mountain hare moults no less than three times per year, going from brown to grey-brown from early to late summer, brown to white in the autumn to midwinter and white to brown again in late winter to early summer. The rate of moulting depends on altitude and temperature and some hares, particularly late season leverets do not go completely white. Populations can be prolific in the right conditions and the series of mild winters in recent years has favoured them as they can have three, possibly four litters, a year. Just occasionally areas containing brown hares overlap with those containing mountain hares but they are easy to tell apart even when the latter are in their brown form as mountain hares have shorter ears, a greyer coat and the black top to the tail is missing.

Dead mountain hares on roads are common in parts of the Highlands especially in the winter, when the hares will use the snow-free roads to get from place to place, and even on remote side roads they fall victim to the occasional vehicle going past. The A9 from Inverness to Perth is notorious for such road casualties, with mountain hares and roe deer being the commonest, and no doubt with the ever present fox only a percentage of casualties are seen. The main predator – apart from man who has treated it as a pest – is the fox, but wildcats and golden eagle will take adults and the leverets are taken by hen harriers, buzzards and stoats. There appears to be a link between the larger numbers of mountain hares in north-east Scotland and the ability for golden eagles to bring off two chicks per year. Eagles start to incubate their eggs when the first one is laid which means that one chick will be older and stronger than the other chick. If there is sufficient food, such as plenty of mountain hares, then both chicks will survive, but if food is short the stronger of the chicks will survive either by killing the weaker chick or starving it to death. It is unusual for both chicks to survive in any of the north-west eagle eyries where mountain hares are much less common.

The best time to see mountain hares is in the winter, particularly where a road runs through moorland or scree over 1,000 feet, as then a vehicle can be used as a hide. They normally feed at night and besides their weight pressing the heather down, their regular pathways are also formed by nibbling the shoots of heather and bilberry. Apart from the

mountain hare there is only one other animal that turns white in winter, namely the stoat, and one bird, the ptarmigan. All three have had problems with the milder winters of recent years, which has left the open moorland devoid of continuous snow. The mountain hares overcome their problem of conspicuousness by sitting near some form of shelter during the day. On moorland, mountain hares will sit by an overhang or shallow hole in the ground where they will dig out their own short burrow up to six feet long. They will also seek out scree slopes but the scree stones must be small enough to run over at speed, while at the same time large enough to dart into recesses in the scree. Several mountain hares will often gather on such a scree slope to sit as scattered individuals rather than as a communal gathering. If a predator such as a golden eagle appears, the animals will then dart into their miniature caves only to come out when the danger has passed. Whether on a scree slope or outside a hole in moorland, the reaction to predators varies. Often the animal will just dart into the hole but sometimes – and it appears to depend on the type of predator – the animal will simply run away, normally uphill.

Although there is some evidence of a decline in re-cent years, there are still some parts of Scotland where mountain hares are common enough to be shot regularly as pests. In some parts, rabbit netting is used to keep both rabbits and hares out from young trees. However in north-west Scotland where numbers are much lower, perhaps because of the much higher rainfall, they are very thin on the ground and tracking in fresh snow indicates that in some areas there may only be one to several hundred acres. Although heather forms 90% of the hares' winter diet they take a range of other food, depending largely on weather. Their liking for short young heather shoots is one of the reasons for their abundance in the north-east as regular burning of moorland to produce this food for red grouse is more than compatible with the needs of the mountain hare. But they will also feed on cotton grass, heath rush, bilberry and deer sedge. Deep or ice-encrusted snow will force the mountain hares to eat even gorse and the bark and twigs of rowan, juniper, birch and willow. They regularly eat their own soft droppings. By night they will move to the lower slopes of the hill to feed and then move back at daybreak, so their territories are long and narrow up and down a hill.

The winter coat can sometimes be conspicuous to predators such as golden eagles.

RED SQUIRREL

Red squirrels playing in old Scots pine woodland perhaps epitomise Scotland's wildlife as much as the red deer or the red grouse for that matter, but all is not well with these attractive mammals that are so at home in trees. Their population in Britain as a whole has always fluctuated, despite the fact that during the second half of the last century they were in all woodland habitats. In contrast the red squirrel was extinct in southern Scotland by the early 18th century and nearly extinct in the Scottish Highlands during the late 18th and early 19th centuries. It was re-introduced into Scotland, mainly with English stock, at ten or more sites between 1772 and 1872. Hence the red squirrel has had a chequered history in Scotland and its future is far from certain, as there is evidence of an unexplained decline in the Highlands. It is not the presence of the grey squirrel as this introduced species has not reached the Highlands, but perhaps the destruction and fragmentation of suitable woodland habitat may well be the main cause of its decline in recent years.

The best way to locate the areas where red squirrels are present is to simply look for stripped cones under the conifers – Scots pine is preferred but they will also use larch and spruce. Once stripped cones have been found then the best chance of seeing them is soon after dawn when they are busy feeding. Crossbills also feed on cones but they merely open them up to extract the seed and do not strip them like the red squirrel. This native mammal – it was one of the last to colonise Britain before it was cut off from the rest of Europe 9,000 years ago – not only takes cones from conifers but also leaf and flower buds, shoots and needles, pollen from male flowers and sappy tissue under bark. But red squirrels also take a variety of other food such as fungi, berries, fruits and even occasionally eggs and young of birds. The cones, however, are the staple diet for nearly all the year and

Red squirrels are secretive, but stripped cones beneath trees can indicate their presence.

the squirrels start on the early green cones.

The summer coat of the adult red squirrel is bright chestnut with orange-brown feet and lower legs. Its ears are tufted and the bushy tail often turns to a pale creamy colour as the summer goes by. In contrast the winter coat is chocolate brown with the fur on the back looking dark grey. The body fur moults in spring and autumn and is unusual in that it moults from the front to the back in spring and from the back to the front in autumn. Most true albino red squirrels, with their characteristic pink eyes, have been seen in Scotland and also very dark brown specimens occasionally occur.

Few predators can catch them in the trees although in Scandinavia they are the main food of the pine marten. Once on the ground, where they forage for acorns and beech mast which they will store for the winter, they are vulnerable to fox, raptors and owls and a few are killed by vehicles when crossing roads.

A specially built drey, with a thick grassy lining, and lodged against the trunk of a tree, will contain three or four young but the male takes no part in the rearing. Older red squirrels have been known to have two broods in a year. The young are born blind and naked but develop rapidly and are fully furred at three weeks old and open their eyes at four weeks. If the drey is disturbed the female will carry the young in her mouth one by one to another drey nearby. The young are most vulnerable during the first few months but winter is the main threat for both the young and the adults. Red squirrels need to build up fat reserves for the winter as, although they may not come out in bad weather, they do not hibernate and they need to feed every few days. If it is a poor cone year and fat reserves are not built up many squirrels will die of starvation or disease and those that do survive may not be fit enough to breed in the following year.

Top: In winter, the red squiirel must eat every few days, often feeding on food stored in the autumn, such as acorns and beech mast.

Bottom: Red squirrels are extremely nimble and alert, and so have very few predators.

GREY SQUIRREL

It took around 30 separate introductions of this North American squirrel between 1876 and 1929 before it became well established, only then to become a pest because of the damage it does to trees. A fierce campaign to control it even led to a bounty being paid for each tail handed in and so the squirrel has been trapped, poisoned and shot. There is, however, a field of thought that efforts to control grey squirrels are a waste of time as creating a vacuum in a population leads to successful influxes of animals that may otherwise have difficulty in finding territories. In some areas pine martens have been introduced to try and kill off local grey squirrel populations as they are able to catch them in the trees, but there has been no proof of success. At present grey squirrels seem to be confined to the central part of Scotland.

Despite the damage these misnamed 'tree rats' do, such as ring barking trees and distorting branches by nibbling shoots, they are attractive and many people watch them in parks and even in their gardens. They have learnt to exploit such places and raid bird tables finding a way around most measures to keep them from the food intended for the birds. Other food is varied, such as sappy tissue under bark, fungi, leaves, flower buds, berries, seeds and even bird eggs. Such foods as acorns, hazel nuts and beech mast are often buried and dug up again in the following spring. Water is generally taken from food or dew but in hot weather they will go to standing water to drink.

Winter dreys are fairly solid structures of twigs and completely round. The grey squirrel will not come out when the weather is bad. Summer dreys are by contrast often merely a platform whereas the breeding drey is more like the winter version. Occasionally they will use a hole in a tree and some have even taken over the chimney-like nest boxes put up for owls. Young squirrels are born in spring or early summer and the average litter of three young ones disperses when they are about ten weeks old. The male takes no part in the rearing. The young can breed after a year. During the campaign against the grey squirrels many dreys were shot out and some contained not squir-

It took a large number of intentional introductions over a 50-year period, before the grey squirrel became established in Britain. They have become quite tame in parks and gardens in central Scotland, and can often be fed by hand. Their winter fur is dense and silver-grey, with a tinge of brown along their back. Their summer coat begins in April or May and turns more brown in colour.

rels, but unfortunately nesting long-eared owls.

Grey squirrels have few predators apart from man because of the difficulty in catching them as they run along branches and jump from tree to tree. They are vulnerable when on the ground and have been taken by fox, raptors and owls. In North America they are classed as a game species and shot for their pelts and meat. Unlike the native red squirrels the grey squirrel can exist in a wide range of habitats and once it colonises an area it slowly increases whilst the red squirrels decrease in numbers. At present the reason for this is a mystery as there are no obvious signs of conflict.

HEDGEHOG

The only spiny mammal in Scotland and the only one that truly hibernates is the hedgehog. Though perhaps one of our best known animals, it is rarely seen because it is nocturnal. Most sightings are of dead hedgehogs as road casualties but they are noisy at night and this can give them away in the garden. Hedgehogs – often called 'urchins' or 'hedgepigs' – may look slow and clumsy but in fact they are surprisingly agile, able to climb and swim well. They are found over most of mainland Scotland although absent from large areas of moorland. Their main habitat is all types of woodland or scrub, including parks and gardens where they are sometimes quite common. Where they occur on islands it is generally

as the result of introductions. The recent spread from such an introduction on South Uist in the Western Isles is causing problems with ground nesting birds as hedgehogs eat their eggs.

Perhaps because they look strange there are one or two myths that have built up round their so-called habits. One myth is that they suck milk from the udders of cows, but no doubt this came about because milk oozing from an udder was drunk by a passing hedgehog. Another story is that hedgehogs roll over and fix fruit such as apples to their spines and carry them off. It would be far more likely for an apple to accidently fall on a passing hedgehog and just stick to the spines! What adds much to the aura of mystery around hedgehogs is their habit of 'self-anointing', when the hedgehogs apply saliva to their spines and the purpose of this is still an unsolved mystery. They are also reputed to kill adders for food and to be un-affected by adder poison. While it may be true that a hedgehog could kill an adder, it is certainly not the case that they are immune to venom, for if an adder bites the soft part of a hedgehog then it can die. White hedgehogs are not uncommon and may seem highly conspicuous but the defence of curling up into a ball of around 5,000 spines puts off most potential predators.

Hibernation starts about November and finishes around Easter but it very much depends on the weather. Late young, now officially called 'hoglets', can get into trouble during hibernation if they have not had time to build up enough fat reserves to last the winter. It has been determined that if a hoglet is not over one pound in weight at the beginning of hiberna-tion, it will not last the winter. Special hibernating boxes can be made for hedgehogs, but they normally prefer to construct a nest of leaves under a bush or garden shed that provides them with protection from the elements. They occasionally wake during the winter and sometimes even build a fresh nest.

The young are born in early summer, but late litters in September are common especially if an early litter has been lost in some way. If a female is dis-turbed in its nest it may eat or abandon newly born young, but older ones are carried away to a new site by the scruff of the neck. Within a few hours of birth, the blind and pink young grow a few white prickly spines and as the young grow older, these are replaced by brown spines growing alongside the white ones. The young start to take solid food at about a month old and the female will lead her family in a procession as they go out to feed. The young are independent after six weeks and the male hedgehog takes no part in the parental care and may not even live nearby.

Apart from being killed by vehicles on the road, hedgehogs face a wide range of hazards, mainly from man's activities. Steep-sided garden ponds are lethal to them and yet a piece of chicken wire in one corner dangling well into the water could easily save many. Hedgehogs often find piles of leaves an ideal resting place and before these are burnt they should be turned over carefully. Slug pellets can cause problems and they should be put well under a stone slab so that the hedgehogs cannot get at them. Hedge strimmers clearing shrubs and hedge bottoms can sometimes severely wound hedgehogs and they should be used with care. Cattle grids also take their toll as the hedgehogs fall in and cannot climb up the steep sides and they are left to starve to death. It should be a legal commitment that all cattle grids have ramps incorporated so that any trapped hedgehogs can climb out safely. One hedgehog was more fortu-nate when a motorist on the A9 near Inverness found the poor animal wandering all over the road at night, with a plastic cup wedged over its face and eyes so that the animal could not see! Few natural predators can tackle the defence mechanism of rolling into a tight ball. Foxes, dogs and badgers seem to get around the problem of the spines and normally leave the remains turned inside out. A few are known to be taken by tawny owls and golden eagles but the main cause of death is on the road and failure to survive the first winter in hibernation.

It has long been traditional to encourage hedgehogs in gardens by putting out bread and milk but this should only be done if they have access to other food, as it gives them diarrhoea. Hedgehogs were formerly eaten, traditionally by gypsies who baked them in mud, but they are now protected under the Wildlife and Countryside Act and cannot be trapped without a licence.

Moles feed mainly on earthworms, which they will occasionally keep in store.

MOLE

Molehills are not entrances to burrows but simply mounds of soil pushed up from the tunnel below and are often the only sign of this secretive species. In the spring the mole builds a nest beneath an extra large molehill, sometimes called a fortress. At one time it was thought that the tunnel system within the fortress was exactly the same in all of them but this was because successive authors writing about the mole kept on copying the same cross-section drawing of a fortress without going out to check. Moles will come up above ground, normally at night, and are occasionally taken by predators. Sometimes the tunnels may be so near the surface that ridges on the ground are formed. At one time these were thought to be made solely by males seeking females, though this was later proved to be wrong. Moles have also been given the name 'mouldywarp' which means 'earth thrower'.

Moles can be found throughout mainland Scotland but are not found on high moorland and hills nor in the Western and Northern Isles. At one time country parishes employed a professional mole catcher to trap and dig up moles and until the 1950s about a million moles were being trapped every year in Britain. Molecatchers would sell the skins which were used for trimming hats and coats. When setting traps, the molecatcher would often use a twig from a willow to mark where the trap had been set. If he did not bother to take the twig out on lifting the traps, the willow wand would frequently take root and grow into a mature tree.

Both sexes are solitary and there are often fights if one mole enters another's tunnel. This active aggression takes place throughout the year apart from a brief mating period in February or March. The female then builds a nest of grass and leaves, about the size of a football, and she will have three or four young in the litter which are born in April or May and become independent after only a month. Most moles have velvety-black fur but colour variations can often occur, with creamy-white, piebald, grey and apricot moles all having been recorded. Some population

Mole hills are not entrances or exits, but the spoil heaps from the underground network of tunnels. Moles sometimes emerge at night, where they can fall prey to owls or are killed by cats and foxes.

studies have made use of the fact that you can put rings on the base of their tails as the base diameter is smaller than the rest of the tail.

Moles have skin glands that make them distasteful to carnivores but large numbers are taken by tawny and barn owls. Their main food source is earthworms and these are sometimes stored alive by biting some of the worms' segments, which makes them immobile. The mole is not blind but has extremely small eyes about the size of a pinhead and it makes its way around in the dark tunnels by using sensitive whiskers and delicate sensors in its nose. There are also sensitive whiskers in its tail so that if it wants to it can run backwards in the tunnel without bumping into anything. To be able to do this, its fur can lie in either direction depending on which way it is going. Moles can swim well and occasionally do so even though it could be avoided.

The wood mouse can readily climb trees and shrubs to seek nuts and berries.

SMALL MAMMALS

Small mammals such as mice and voles are, despite their size, a very important link in the food chain for a wide variety of animals and birds. We may talk about the profound changes man has made in his manipulation of the landscape such as over-burning, over-grazing, drainage, pollution, poisoning, shooting, the destruction of forests and the consequent effect this has had on deer, birds of prey, fish and plant life, but such changes have also adversely affected those small mammals that we may see only as they dart across the road in front of a vehicle. An obvious example of this effect is the attractive sight of a kestrel hovering over the roadside verges. It could be said that it is a good thing that wildlife are utilising such artificial habitats but it is in fact a sad reflection that the reason the kestrel is there is that such long strips of grass provide some of the only habitat left for the mice and voles that the kestrel will be hunting. Intensification of land-use practices have drastically reduced the hunting, feeding and breeding ground for our small mammals and as food availability is one of the main controlling factors of numbers and distribution then those species feeding on small mammals, such as kestrels and weasels, are bound to also have been adversely affected. Predators of mice, voles and shrews include fox, stoat, badger, pine marten, wildcat, long-eared owl, barn owl, short-eared owl and tawny owl.

The wood mouse – also known as the long-tailed field mouse – is found throughout Scotland and is absent only from very small islands. It may well be the most widespread and abundant mammal in Scotland. It seems likely that some of the northern islands have had wood mice accidently brought in by man from Scandinavia. A great deal of work has been carried out to isolate island forms and even sub-species, with the larger forms on islands such as Rum, Lewis and St Kilda in particular. On St Kilda it inhabits the old houses and the cleitean, where the wood mouse is a separate sub-species with a specific name of 'hirtensis' after the main island of Hirta. Wood mice are very agile and readily climb trees where

The harvest mouse faces an uncertain future in Scotland.

they will go into bird nest boxes and they frequently store hips and haws in old small birds' nests. They are rarely seen as they are mainly active at night.

Little is known about the present distribution of the house mouse but there is evidence that it has declined in the houses and outhouses of many areas, where its place has been taken by the wood mouse. Such is the case on St Kilda as the house mouse, probably brought in by the Vikings, lived happily in the village on Hirta until the St Kildans left in 1930. Then food became short, the houses were no longer heated and there were cats left to fend for themselves. The larger wood mouse moved into the houses and within 18 months of the evacuation the St Kilda house mouse was no more. It had been another sub-species for St Kilda.

Unlike other mice the house mouse has a strong smell and greasy fur and it taints houses where it lives. They are also very difficult to eliminate as they multiply rapidly. They build their nests from any material that can be torn up, such as newspapers, grass or

Top: The house mouse can be a destructive visitor in the home, often chewing through electric cables.

Bottom: The St Kildan wood mouse still survives today, while the St Kilda house mouse soon became extinct after the people of St Kilda were evacuated from the islands.

In northern Scotland, the water vole is often more black than brown.

sacking. Females can have from five to ten litters a year with five or six young in each. The young leave the nest when they are three weeks old and the females amongst these are ready to breed after another three weeks.

By far the rarest of the small mammals in Scotland, and one of the world's smallest rodents at around a quarter of an ounce, is the harvest mouse that can still be found in the south-east of the country. At one time it was more widespread but has had to learn to cope with the intensification of farming and, as it builds its round nest well off the ground, it now resorts to hedges, reeds, brambles or scrubby areas. The nest is made of live grass blades and is interwoven between stems and is about the size of a tennis ball – tennis balls with a hole in the side have actually been used as harvest mice 'nest boxes'. They eat a variety of plants and sometimes insects and move through vegetation with great agility, using the tail as a fifth limb. In good weather they can have several litters of up to six young

in a season. They become independent at 16 days and are grey-brown as opposed to the reddish-brown upper parts of the adults. Harvest mice are active by day and night so their predators include both day and night hunters. The easiest way to locate the harvest mouse is to wait until the vegetation dies down in the autumn when their nests are more conspicuous. For the winter they moult to a dark brown coat and they resort to low vegetation but will sometimes enter buildings.

The water vole (the famous Ratty of *Wind in the Willows* was in fact a water vole) is often called 'water rat', but in fact it is only distantly related to the rat. They live on and in the waterside banks and feed almost entirely on waterside plants. They are fairly well distributed in southern and eastern Scotland but have a patchy distribution in the north. There is a northern colour form of the water vole that is black all over rather than the usual brown fur but it is not a sub-species as such. The voles form a system of

After the wood mouse, the bank vole is probably Scotland's most abundant rodent.

burrows in banks and the entrance may be above or below the water. They are often active by day and if a vole hole is found or a feeding area indicated by browsed plants, then slices of apple placed nearby will often attract them and they can then be watched feeding in the open. Occasionally brown rats are found in similar places but the blunt nose of the vole together with long fur that hides the ears and the shorter tail readily distinguishes them.

The second most abundant small mammal is almost certainly the bank vole and it much prefers to live in dense cover such as bramble thickets, hedgerows and scrub where it is active by day. They will run along a network of well-worn tunnels in vegetation in their search for food but live-trapping indicates that they generally stay within 50 yards of their nests. Hazelnuts are frequently eaten and the tell-tale nibbled hole in the side gives the culprit away. They will also store food such as fruit, berries, nuts and seed underground. Despite being only about one ounce in weight they climb readily and are well-balanced as they reach out for berries or fruits. Although a vole born early in a mild year can have its own family after a few weeks, mortality is high at 50% within the first few months. Predation and cold weather are the two main reasons for this high mortality rate.

Vole plagues are generally associated with the field vole and its population cycles of peaks and troughs. Accordingly it has a very large number of predators with nine other species of mammals preying on it, eight species of bird and one reptile – the adder. Birds such as the short-eared owl are reputed to be able to forecast vole plagues and they lay a larger number of eggs so that they can bring off more young with such an abundance of food. However after a plague year, which occurs in cycles of three to five years, the number of voles drops drastically. No one really knows why, though some theories suggest overcrowding or aggression. Field voles may form as much as 90% of a barn owl's diet and it may well be that one of the many reasons for the decline of the barn owl in Scotland has been the decreasing amount of habitat suitable for field voles.

High-pitched squeaking in a wood pile or hedge

The recent decline of the field vole may well contribute to that of the barn owl.

bottom is likely to be from a common shrew screaming at another shrew which has invaded its territory. Such fights happen frequently as a shrew will not tolerate any other shrew in its territory unless it is breeding time. Shrews are so small that they must eat constantly and so they are nearly always on the move by day and by night. The reason they are so seldom seen is that they spend three quarters of their time underground. One method of studying small mammals is by live trapping but if shrews are likely to be caught then a small hole must be left in the side of the trap so that shrews can get out whilst retaining voles and mice. If you want to catch shrews to study then a licence under the Wildlife and Countryside Act is required and the trap should be checked every two hours, as otherwise the shrew will die of starvation. Its distribution in parts of northern Scotland may be limited by the absence of earthworms on moorland, as earthworms

The water shrew (Top) and common shrew (Bottom) both have characteristic long noses.

The pygmy shrew can weigh only a fraction of an ounce.

are one of the common shrew's main foods.

The smallest mammal in Scotland – in Britain for that matter – is the pygmy shrew, so called because it is only a fraction of an ounce in weight. Like the common shrew it can be recognised – if by chance it is seen – by its long, pointed nose. It takes a wider range of food than the common shrew and as it is not therefore affected by the absence of earthworms, it is more widely distributed. The pigmy shrew will starve to death if it does not eat for around two hours and it is so tiny that it is near to the minimum limit at which a warm-blooded animal can exist, as its body nearly loses heat too rapidly for it to survive. This means that it forages virtually all the time, but despite its size it is very hardy and providing food is available it can survive a very wide range of conditions and has been seen on some of the highest hilltops in Scotland. They are preyed on by barn owls which do not seem to mind the foul-tasting glands in the shrew's skin, which deter other predators such as weasels and stoats. Cats frequently catch shrews but will not eat them because of their taste.

Although it has a black coat and feeds in water, the water shrew can be distinguished from the black northern form of the water vole because it has the characteristic long nose of the shrew family. Although they will hunt in the water for fish, tadpoles and frogs, they also spend a great deal of time away from water. If they catch something in the water it is usually dragged to the shore and then eaten. The mild poison in the shrew's saliva may help to subdue larger prey such as frogs. If seen swimming under water these shrews may appear silvery because of air trapped under their fur and consequently must keep paddling as this trapped air makes them buoyant. Predators include large trout, pike and owls and in some areas water shrew numbers are controlled by mink.

Rats on islands are a regular nightmare for

Black rats are reputed to have brought in the plague of the Middle Ages.

conservationists as they will take eggs and young of ground nesting birds. In contrast the black rat, which reached Britain around the 11th century, is reputed to have brought with it the plague that killed more than a third of Britain's population in the 14th century, although new ideas suggest it may not have been the plague that decimated the population. They can survive on islands by scavenging the shore lines until sea birds nest again in the spring. They are very agile at climbing ropes, and cones are often put on mooring ropes of ships which are positioned so that they stop rats getting on the ships but allow them to get off. In the past both black and brown rats were stopped from getting to grain at food stores by large mushroom-shaped 'staddle stones' which entirely supported the buildings.

The black rat was once widespread but when the more aggressive and adaptable brown rat came to Britain 200 years ago, the black rat almost disappeared, though it is still found at some ports or old towns on the coast. The brown rat is reputed to have arrived from Russia in the 18th century and part of its success results from the very wide range of habitats it occupies and the very wide range of food it will take. It is equally at home raiding a bird table in the garden whilst at another extreme it raids the shearwater burrows high on the hills of Rum in the Inner Isles. If food and shelter are ideally available a female brown rat can have five litters a year, totalling 50 young. They will thrive in farm outbuildings and their love of sewer systems means they readily transmit diseases. Despite persecution from trapping, poisoning and the use of dogs and cats they still flourish. During the summer they often live out in the countryside but with the start of the first frosts they will head for buildings or sewers.

Brown rats can create havoc on islands for nesting bird colonies.

Common seals are not as common as grey seals in Scotland.

COMMON SEAL

During the last two decades two major incidents have affected seals in Scotland. The first was the six-year plan from the Scottish Office which intended to severely cull Scotland's grey seal numbers in the so-called interests of the white fish fisheries, but public outcry stopped this. However, there was nothing any-one could do about the second major incident when common seals around the Scottish coast began to die from a virus similar to canine distemper that lowers the seal's immunity to infections. In 1988 over 6% of the estimated Scottish population of around 18,000 were found dead; the percentage was much lower than in England where some populations were reduced by 21% whilst some of the continental popu-lation were reduced by between 75–80%. It was esti-mated that the total number of common seals found dead on continental shores was over 14,000. It would be very difficult to even begin to estimate just how many seals died and were never located.

Such epidemics have happened before and were documented in 1770, 1836 and 1869, but what was remarkable in 1988 was the combining of interests of organisations and individuals to supply information and collect samples from seals for analysis. A 'Save Our Seals' campaign was launched by the Scottish Wildlife Trust and a seals campaign officer was appointed in September 1988 with a 24-hour answer-phone in operation, with financial help from the Vincent Wildlife Trust and the Nature Conservancy Council. The first dead seals were washed ashore in Scotland in mid-August.

Common seals are fairly easy to see on the Scottish coasts although they do not form such large groups as the grey seal. They prefer to haul-out on sand or mud banks in firths or estuaries rather than the rocky shore used by grey seals. Despite a range of differences between the common seal and the grey seal it is not al-ways easy to tell one from the other. The common seal has a short muzzle with a rounded head and a uniform spotted coat that is paler underneath. The grey seal's muzzle is elongated and the bull's resembles a 'Roman nose' profile. The grey seal's coat is variable

Unlike the grey seal, common seal pups can swim soon after birth. Adult common seals are far less sociable than grey seals.

and patchy with the bull's fairly dark overall. They are much larger than the common seal but these are all relative indicators and there is perhaps only one sure way to tell them apart, but it involves getting close enough to see the nostrils. This may not be too difficult as seals are so inquisitive, but the distin-guishing feature between the two is the 'V' shape of the common seal's nostrils, as opposed to the parallel nostrils of the grey seal.

Compared with the weaning period on rocky coastlines of the grey seal, the common seal's breed-ing behaviour is striking. In late May the cow has accumulated a thick layer of blubber and is heavily pregnant and will start to come ashore far more than usual. In June the pup is born and remarkably can swim within the first hour of birth. This has several advantages over the grey seal as there is less risk from predators, particularly man, because the cow and pup can simply swim away. The grey seal pups are much

more vulnerable at birth. This is why grey seals must seek out remote, undisturbed islands for their rookeries whilst the common seal can freely pup in firths and estuaries quite close to man and his activities. The conspicuous white coat that characterises the grey seal pup is not often seen on the common seal pup as it is moulted before it is born and the pup comes into the world in its first adult coat.

Making contact with common seals is fairly easy as they will often follow boats such as ferries. When the Kessock Ferry was plying between Inverness and Kessock in the Beauly Firth both common and grey seals regularly followed the boat. Boat trips off the west coast of Scotland are often organised to see both birds and the common and grey seals. The roads running alongside firths and estuaries up the north-east coast are often good vantage points to see common seals. Unlike the grey seal that must move some distance to its sometimes huge rookeries, the common seal appears to be fairly resident, partly because it can feed and breed in the same areas. However, by not collecting together in large numbers like the grey seals, they are much more difficult to count and if you see some hauled-out on a sand bank you never really know how many are still in the water hidden from view. Tagging and other surveys in local populations have given much higher estimates than previous figures suggested, so it could well be that the total Scottish population of common seals is much greater than the estimated 18,000.

Common seals may look ungainly on land, but are most graceful in water.

GREY SEAL

One of the wildlife spectacles of Europe is a large breeding rookery of grey seals and the two largest rookeries are found on remote islands off the north and western seaboards of Scotland. The largest is on the archipelago of the Monach Isles, a few kilometres west of Benbecula in the Western Isles. The 12,000 adults that gather there in October produce nearly 6,000 pups each year and it is generally considered to be the largest grey seal rookery in the world. North Rona, an even more remote island off Cape Wrath in north-west Sutherland, also has 12,000 adults but the pup production at 2,000 per year is much lower. The British population of grey seals has been estimated at around 100,000 which is about half of the world population and around 90% of these are found in Scotland.

On North Rona the grey seals will drag themselves out some distance from the sea, up to the 200-foot contour over grassland and rocks. The bulls establish small territories which they actively defend. The Sea Mammal Research Unit has carried out many years of study on this island, made famous by the writings of Sir Frank Fraser Darling. In contrast the Monach Isles are low-lying shell sand islands with the grey seals scattered around the edges of the island on small beaches with rocks and sand. Few studies have been made of this huge rookery but counts in recent years have been made by using aerial photography.

However it is not necessary to try and reach these remote islands to see grey seals, as outside the breeding season they are found throughout Scotland's coastal waters, from harbour to rivers and sea loch. They frequently haul-out on rocks or mud and sand banks in the same way as common seals but 'haul-out' is generally a misnomer because all the seals actually do is sit on the first exposed piece of rock or mud and then the ebbing tide may well leave them some distance from the water. Unlike the common seal it is not possible for grey seals to pup at such sites as the pups must be fed on land for three weeks before they are deserted by the cows and eventually take to the sea.

The history of the grey seals in the last few centuries is a varied one. In the 19th century large numbers were taken for their oil, meat and skins. This was followed by commercial hunting for their skins and by the time legislation came out in the form of the Grey Seal Protection Act of 1914 it was estimated that there might only be 500 grey seals left in Britain. This was almost certainly an underestimate but it suggests how serious the situation had become, though few observers then could have foretold the dramatic change in the numbers of today.

The reasons for the increase of grey seals is probably a combination of many factors but certainly the legislation leading to a decrease in shooting them for oil, meat or commercially for skins, has aided their recovery. What seems fundamental to the change, however, is the number of islands, such as the Monach Isles and North Rona, which have become less populated in recent years, leaving behind ideal and undisturbed breeding grounds for grey seals. The Monach Isles are a prime example. When people left the island in 1949, the grey seals moved in and by the early 1960s a small number were breeding. However by 1985 the rookery had grown so large that it was producing over 4,000 young per year and even with nearly 6,000 in 1989, there is still plenty of room for an even further increase.

On many of the islands supporting breeding rookeries of grey seals, there are also large and chaotic-looking sea bird colonies. But by the time the grey seals are pupping – varying from early October on the Monach Isles to as late as November on the Isle of May – most of the sea birds have gone. They are replaced by the apparent chaos of the seal colony, although as with the sea birds it is not as chaotic as it first appears. The first to return to the rookery are the fat and pregnant cows. They visit the site many times as if looking it over for a suitable place to pup. About a day prior to the birth, the cow will stay on land and eventually the pup is born to the screams of gulls as they squabble over the membrane and then the placenta. The cow will argue with anything that comes too close – from the gulls to other seals. There may be a link between the increase in seal numbers

and an increase in greater black-backed gulls because of the food available at such colonies, whether from afterbirth, dead pups, or for that matter dead adults.

As soon as the first pups are born the bulls come ashore and compete for the best breeding sites, although it does appear that the cows dictate where they will stay rather than the bulls. The bull protects his territory within which he may have from two to ten females. Although this is often achieved by aggressive posture rather than physical contact, there are fights and these can often be savage, especially with the larger bulls, and much blood can flow from wounds. Many of the older bulls show scars from such battles. The pup is by then nearly weaned and soon after mating, the cow leaves for the sea again, abandoning the pup to fend for itself.

Grey seals are kept in captivity in a number of places but they are best seen in their natural element and although they appear along the coast there are many boat trips out of harbours that are partly designed to look at both seals and sea birds. The seals are very curious and will often allow a boat to get very close to rocks where they are resting and even when a seal dives into the sea it will keep bobbing up around the boat affording excellent views. However if you are on land and choose to look at a haul-out never disturb the seals and never get too close as the adults can move very quickly and are very strong with powerful jaws and teeth. Pups may look innocent and cuddly but they too are strong with very sharp teeth!

Grey seals spend most of the time in the water and although we may see them along the coast, much of their lives in the seas are still shrouded in mystery. This, combined with their large size, has given them a distinctive place in folklore and in particular the selkie legends. Selkies were seals that could leave the sea in human form with various bad intentions. In one version the selkie appears in the form of a female who entices young men into the sea, by making them fall in love with her. Another popular version is that the selkie is a male who makes a woman pregnant only to return later to claim the child and take it into the sea as a baby seal. Their lovely moaning sounds are reminiscent of a sensitive lowing of cattle but this has been linked to the sirens who entice ships and people onto dangerous rocks.

Man has persecuted grey seals for a number of reasons in the past, but more recently because of the controversy over seal numbers and their effect on fish stocks. This came to a head in the late 1970s when there was great pressure against grey seals from a wide variety of fishing interests – from anglers after

Grey seals gather in large breeding rookeries to bear pups and mate.

salmon in the traditional fly-fishing methods to commercial fishing for a variety of species. Eventually the department of Agriculture and Fisheries for Scotland decided to look into culling the seals at their breeding rookeries. But public concern at these plans overcame even the pressure from the fishing bodies and the culls were eventually called off. The current number of grey seals have no doubt affected fish stocks, even in the seas, but the situation has been aggravated by our own overfishing of the resource.

Many grey seals are illegally shot in river systems by keepers and around fish cages in sea lochs, but compared with the large numbers of seals around the coast, such culls have little effect on the population. Mortality at the breeding grounds and just after the pup goes to sea is far more influential in determining the increase of numbers. Mortality at rookeries is normally up to 25%, from such things as stillbirths and

squashing at overcrowded sites. The pup's first weeks at sea are also critical when it must learn to fend for itself. The overall result is that only about a third of pups ever reach their first birthday. Natural predators are few around Scotland's coasts although killer whales will take them as food.

In the late 1980s a close watch was kept on both the North Rona and Monach Isles' rookeries (they are both National Nature Reserves) when fears were voiced that the virus that killed so many common seals might decimate the greater concentration of grey seals, but fortunately this did not occur. Grey seals can cause erosion problems at their rookeries particularly where the seals are on machair or sand dune systems. However at the Monach Isles' site where the grey seals utilise pockets of bare sand, sometimes some distance from the sea, it appears that these areas are caused by wind erosion rather than by the seals themselves.

Top: The smooth or common newt is the second most widespread newt in Scotland.

Bottom: The palmate newt is by far the most common newt in Scotland, especially in upland areas where it will breed in bog pools.

A silvery streak on the tail is a distinguishing feature of the male great crested newt.

AMPHIBIANS & REPTILES

Very few studies have been carried out on the amphibians and reptiles in Scotland and the result is a lack of knowledge not only of their behaviour but also of their distribution. One of the reasons for this is that whilst birds and to a lesser extent plants and mammals are fairly well covered by naturalists, they are very few observers seriously interested in amphibians and reptiles.

Until the last decade the existing information on the three species of newts in the Highlands was confusing as the few authors that actually mentioned newts seemed merely to copy information given by the authors Harvie-Brown and Buckley, writing at the end of the last century. The smooth newt was said to be

the commonest, the palmate newt was widespread and the great crested newt was, as the books say, 'found in deep water'. This information has been repeated by several authors since and was even repeated in two books published in the last few years. At one time the newt was called an 'eft' which later became an 'ewt' but this was difficult to say and it eventually became a 'newt'.

One of the problems of identifying newts is that the female palmate and smooth newts are almost identical. However, an adult male palmate newt has a thin black filament at the end of its tail. The great crested newt is twice the size of the others and, as its other name of 'warty newt' suggests, it has a lumpy-looking skin. In the last 15 years, hundreds of lochs in the

A yellow streak down the back will help to identify the natterjack toad.

Many predators will not bother toads because of their unpleasant skin secretion.

Highlands have been surveyed and it is extremely likely that only the palmate newt is native to the north. Only three colonies of smooth newts and three colonies of great crested newts were located. Two of the smooth newt colonies are known to have been deliberate introductions and it seems likely that the other colony plus the three colonies of great crested newts were also introduced. Such introductions were a regular event after World War II when many pet shops stocked amphibians and reptiles. When owners became tired of them they turned them out into the nearest pond or others simply escaped.

Newts are not active by day because they have so many predators and they spend their time in dense weed cover or under stones. By night, however, they are very active and at a good breeding colony in the spring, a torch will reveal hundreds of newts wandering around looking for food or a mate. If you have no

choice but to look in daylight then try turning over stones in the shallow water at the margin of a loch or lochan where the water temperature is slightly warmer. One important point to remember is that the stones should be replaced in exactly the same spot as there may be many creatures sheltering underneath. Most newts will spend the winter away from water in crevices under stones or in walls but there is one known colony of palmate newts that spends the winter underwater in a loch. This is Loch Sgaorishal at 600 feet on the island of Rum.

Every spring there are reports of large numbers of frogs killed by traffic on certain stretches of roads in Scotland but it is rare that frogs are involved. It is normally common toads that are sometimes unfortunately killed in their hundreds. It is unlikely that frogs are involved because they spend the winter in various places around their breeding site and can actually

Common frogs will gather at times in large breeding colonies.

Slow-worms like to bask in partial, but not direct sunlight.

spend the time underwater. In contrast, nearly all the toads from a colony will 'migrate' in one direction from a pond and will hibernate in roughly the same area. If cool weather is followed by warm sultry weather with some rain, all the toads leave their winter quarters and head off in the same direction en masse and where they cross a road the accidents can happen.

Toads are generally found in deeper water than frogs and the two are rarely found at the same breeding site. The spawn of the two species is easily told apart because that of the toad is a long line of eggs wrapped around vegetation whereas that of the frog is laid in dense clumps. Fortunately frogs are still common and widespread in Scotland unlike in southern England, where because of changes in habitat they are almost restricted to garden ponds! One of the main ways to differentiate between frogs and toads is the fact that at all ages – even froglets – there is a dark patch behind the eye which is absent in the toad. Most predators such as weasels and stoats tend to leave toads alone as when threatened toads will secrete an odorous poison onto their skin that is distasteful to the predator. Other predators include crows, herons,

ducks, rats, foxes and cats.

The natterjack or 'running toad' is only found in the extreme south-west of Scotland and it is easily identified by the yellow line down the middle of its back. It occurs mainly in sand dunes where it lays its spawn in pools. This has been its downfall as these pools often dry up during the summer and the tadpoles die. Because of the national decline in numbers, the natterjack toad is given special protection under the Wildlife and Countryside Act and great efforts have been made to manage habitats for it.

There is only one true lizard in Scotland, the common or viviparous lizard that, as its name suggests, is a live bearer. Most visual contacts with lizards are after a rustle in the herbage and a brief glimpse is had of a small brownish-grey lizard darting into cover. Early in the summer, however, they spend a great deal of time basking in the sunshine and they can be watched for some time. They are found in a wide variety of habitats from moorland to sand dunes and grassland to scrub-covered hillsides or banks. Their predators include kestrels, adders and rats. If a predator grabs the lizard by its tail the end of the tail drops off

The tail of the common lizard will sometimes drop off if seized.

enabling the lizard to escape. The tail eventually re-grows but never to its former length and shape and many lizards show the result of such predation with blunt tips to their tails.

As the grass snake is not found in Scotland the only two snake-like animals are the adder and the slow-worm. In general the zig-zag pattern on the adder and its larger size readily separates it from the fairly uniform colour of the brownish slow-worm. The adder is a true snake whereas the slow-worm is of the lizard family – albeit a legless lizard. Adders are found in a variety of dry or damp places such as moorland or scrub-covered hillsides but they are fairly localised. The adder is the only poisonous snake in Britain and whilst medical treatment is required it is rarely serious unless someone is allergic to the venom. Anti-venom serum is no longer given as some people were reacting more to the serum than the poison and patients are now normally given antibiotics. Although there is evidence of a decline in the adder in Scotland – decrease in habitat and pointless predation by man are the reasons – there was great debate as to whether it should be given protection under the Wildlife and Countryside Act and it was not fully protected until 1991. The other name for the adder is 'viper' after its viviparous nature of being a live bearer.

The slow-worm looks like an overgrown worm and it moves slowly, hence its name, and many are still killed when they are mistaken for snakes. They are widespread throughout Scotland but are not often seen as they rarely bask in the open. Both males and females are brownish although the females normally have dark stripes. Older male slow-worms have blueish spots on back and flanks, especially near the head, but the reason for their presence is a mystery. Each young slow-worm is born in a membranous egg that it breaks open within seconds and the young are golden yellow above and black below. Mating takes place in April and May and at this time males often fight, biting each other, especially on and about the head whilst their bodies intertwine. The small young have many predators from frogs to birds whilst adult slow-worms fall to adders, rats, kestrels and even buzzards. Being in the lizard family means that like the common lizard, if it is seized by the tail, the tail will drop off and partly regrow. Hibernation is under-ground and lasts from October to March.

If an adder's victim does not die quickly, the snake will often follow it.

SELECTED BIBLIOGRAPHY

Corbet, G B & Harris, S, *The Handbook of British Mammals*. 3rd Edition, Blackwells, 1991.

Corbet, G B, *The Terrestrial Mammals of Western Europe*. Foulis, 1966.

Reader's Digest, *Reader's Digest Guide to the Animals of Britain*. Reader's Digest, 1984.

Stephen, D, *Highland Animals*. Highlands and Islands Development Board, 1974.

Thompson, F, *A Scottish Bestiary*. Molendinar Press, 1978

Whitehead, G K, *Wild Goats of Great Britain and Ireland*. David & Charles, 1972.

There are currently a range of monographs on some of the animals included in this book and the following publishers should be consulted:

Colin Baxter's Wildlife Series: A series of informative books illustrated with full colour photography.

Shire Natural History Series: A series of booklets that are informative and enjoyable.

Whittet Books: A series of books that are not only informative and enjoyable, but contain amusing cartoon illustrations.

Christopher Helm Mammals Series: A series of more scientific and comprehensive books.

BIOGRAPHICAL NOTE

Ray Collier has been a professional conservationist since 1961 and from 1977 has been Chief Warden with the Nature Conservancy Council in the Highlands. He has studied mammals, butterflies, dragonflies, amphibians, reptiles and birds, specialising in habitat and species management. His other interests include photography, writing, poetry, rugby and dachshunds. He lives with his wife in an old farmhouse in the Inverness-shire Highlands.